T0215839

Lecture Notes in Computer Science   11360

Commenced Publication in 1973
Founding and Former Series Editors:
Gerhard Goos, Juris Hartmanis, and Jan van Leeuwen

More information about this series at http://www.springer.com/series/8637

Abdelkader Hameurlain · Roland Wagner
Franck Morvan · Lynda Tamine (Eds.)

Transactions on Large-Scale Data- and Knowledge- Centered Systems XL

Springer

Editors-in-Chief
Abdelkader Hameurlain
IRIT, Paul Sabatier University
Toulouse, France

Roland Wagner
FAW, University of Linz
Linz, Austria

Guest Editors
Franck Morvan
IRIT, Paul Sabatier University
Toulouse, France

Lynda Tamine
IRIT, Paul Sabatier University
Toulouse, France

ISSN 0302-9743 ISSN 1611-3349 (electronic)
Lecture Notes in Computer Science
ISSN 1869-1994 ISSN 2510-4942 (electronic)
Transactions on Large-Scale Data- and Knowledge-Centered Systems
ISBN 978-3-662-58663-1 ISBN 978-3-662-58664-8 (eBook)
https://doi.org/10.1007/978-3-662-58664-8

Library of Congress Control Number: 2018966822

This Springer imprint is published by the registered company Springer-Verlag GmbH, DE
part of Springer Nature
The registered company address is: Heidelberger Platz 3, 14197 Berlin, Germany

Preface

This volume contains five fully revised selected regular papers, covering a wide range of very hot topics in the fields of social networks, data stream systems, and linked data. These include personalized social query expansion approaches, continuous query on social media streams, elastic processing systems, and semantic interoperability for smart grids and NoSQL environments. We would like to sincerely thank the editorial board and the external reviewers for their thorough reviews of the submitted papers and ensuring the high quality of this volume.

Special thanks go to Gabriela Wagner for her availability and her valuable work in the realization of this TLDKS volume.

October 2018

Abdelkader Hameurlain
Franck Morvan
Lynda Tamine
Roland Wagner

Organization

Editorial Board

Reza Akbarinia	Inria, France
Bernd Amann	LIP6 – UPMC, France
Dagmar Auer	FAW, Austria
Djamal Benslimane	Lyon 1 University, France
Stéphane Bressan	National University of Singapore, Singapore
Mirel Cosulschi	University of Craiova, Romania
Dirk Draheim	Tallin University of Technology, Estonia
Johann Eder	Alpen Adria University Klagenfurt, Austria
Anastasios Gounaris	Aristotle University of Thessaloniki, Greece
Theo Härder	Technical University of Kaiserslautern, Germany
Sergio Ilarri	University of Zaragoza, Spain
Petar Jovanovic	Universitat Politècnica de Catalunya, BarcelonaTech, Spain
Dieter Kranzlmüller	Ludwig-Maximilians-Universität München, Germany
Philippe Lamarre	INSA Lyon, France
Lenka Lhotská	Technical University of Prague, Czech Republic
Vladimir Marik	Technical University of Prague, Czech Republic
Jorge Martinez Gil	Software Competence Center Hagenberg, Austria
Franck Morvan	Paul Sabatier University, IRIT, France
Torben Bach Pedersen	Aalborg University, Denmark
Günther Pernul	University of Regensburg, Germany
Soror Sahri	LIPADE, Descartes Paris University, France
A Min Tjoa	Vienna University of Technology, Austria
Shaoyi Yin	Paul Sabatier University, France
Osmar Zaiane	University of Alberta, Edmonton, Cananda

External Reviewers

José María De Fuentes	Universidad Carlos III de Madrid, Spain
Tamer Elsayed	Qatar University, Qatar
Evangelos Kanoulas	University of Amsterdam, The Netherlands
Riad Mokadem	Paul Sabatier University, France
Eric Pardede	La Trobe University, Australia

Contents

Personalized Social Query Expansion Using Social Annotations

Mohamed Reda Bouadjenek[1], Hakim Hacid[2(✉)], and Mokrane Bouzeghoub[3]

[1] Department of Mechanical and Industrial Engineering, University of Toronto,
Toronto, Canada
`mrb@mie.utoronto.ca`
[2] Zayed University, Dubai, United Arab Emirates
`hakim.hacid@zu.ac.ae`
[3] University of Versailles, Versailles, France
`mokrane.bouzeghoub@uvsq.fr`

Abstract. Query expansion is a query pre-processing technique that adds to a given query, terms that are likely to occur in relevant documents in order to improve information retrieval accuracy. A key problem to solve is *"how to identify the terms to be added to a query?"* While considering social tagging systems as a data source, we propose an approach that selects terms based on (i) the semantic similarity between tags composing a query, (ii) a social proximity between the query and the user for a personalized expansion, and (iii) a strategy for expanding, on the fly, user queries. We demonstrate the effectiveness of our approach by an intensive evaluation on three large public datasets crawled from delicious, Flickr, and CiteULike. We show that the expanded queries built by our method provide more accurate results as compared to the initial queries, by increasing the MAP in a range of 10 to 16% on the three datasets. We also compare our method to three state of the art baselines, and we show that our query expansion method allows significant improvement in the MAP, with a boost in a range between 5 to 18%.

Keywords: Personalization · Social Information Retrieval
Social networks · Query expansion

CR Subject Classification: H.3.3 [Information Systems]: Information
Storage and Retrieval · Information Search and Retrieval

1 Introduction

Web 2.0 has strengthened end-users position in the Web through their integration in the heart of the content generation ecosystem. This has been made possible mainly through the availability of tools such as social networks, social bookmarking systems, social news sites, etc., impacting the way information is produced, processed, and consumed by both humans and machines. As a result,

© Springer-Verlag GmbH Germany, part of Springer Nature 2019
A. Hameurlain et al. (Eds.): TLDKS XL, LNCS 11360, pp. 1–25, 2019.
https://doi.org/10.1007/978-3-662-58664-8_1

on the one hand, the user is no longer able to digest the large quantity of information he has access to and is generally overwhelmed by it. On the other hand, most of popular Information Retrieval (IR) systems lack in offering efficient personalization techniques, which provide users only with the necessary information that fulfill their needs. Two types of constraints make the situation more complex: *information-dependent* constraints and *user-dependent* constraints. The first class of constraints includes (i) the large scale due to the continuous activities of users and their ability to generate new content, (ii) information diversity or heterogeneity, since different types of media are used to communicate, e.g., text, image, video, etc. (iii) versatility, since information is dynamic and is continuously updated (confirmed, contradicted, etc.), (iv) its disparity, since it can be in different places, and as a result (v) the variation in the quality of information. The second class of constraints is mainly related to users' diversity and the high dynamics in their profiles.

To improve the IR process and reduce the amount of irrelevant documents, there are mainly three possible improvement tracks: (i) query reformulation using extra knowledge, i.e., expansion or refinement of the user query, (ii) post filtering or re-ranking of the retrieved documents (based on the user profile or context), and (iii) improvement of the IR model, i.e., reengineering of the IR process to integrate contextual information and relevant ranking functions. In this paper, we focus on query reformulation, especially on personalized query expansion for personalized search, i.e., personalizing the reformulation of queries.

Query expansion consists of enriching the user's initial query with additional information so that the IR system may propose suitable results that better satisfy user's needs [14,15,19]. We explore the possibility of using the data available in social networks, and more precisely data of social bookmarking systems, as a source of explicit feedback information. These latter enable users to freely add, annotate, edit, and share bookmarks of web resources, e.g., web pages. Basically, we propose an approach which reuses the users vocabulary (the terms used to annotate web pages) in order to expand their queries in a personalized way and thus, increase their satisfaction regarding the quality of search. Exploiting social knowledge for improving web search has a number of advantages:

- Feedback information in social networks is provided directly by the user, so users interests accurate information can be harvested as people actively express their opinions on social platforms. Thus, this user interest can be easily modeled to provide personalized services.
- A huge amount of social information is published and available with the agreement of the publishers. Exploiting these information should not violate user privacy, in particular social tagging information, which doesn't contain sensitive information about users.
- Finally, social resources are often publicly accessible, as most of social networks provide APIs to access their data (even if often, a contract must be established before any use of the data).

Our approach in this work[1] consists of three main steps: (i) determining similar and related tags to a given query term through their co-occurrence over resources and users, (ii) constructing a profile of the query issuer based on his tagging activities, which is maintained and used to compute expansions, and finally, (iii) expanding the query terms, where each term is enriched with the most interesting tags based on their similarities and their interest to the user.

The problem we are tackling in this paper is strongly related to personalization since we want to expand queries in a personalized way and consequently propose adapted search results. Personalization allows to differentiate between individuals by emphasizing on their specific domains of interest and their preferences. It is a key point in IR and its demand is constantly increasing by various users for adapting their results [3]. Several techniques exist to provide personalized services among which the user profiling. The user profile is a collection of personal information associated to a specific user that enables to capture his interests. Details of how we model user profiles are given in Sects. 2 and 3.1.4.

The main contributions of this work can be summarized as follows:

1. We propose an approach in which we use social knowledge as explicit feedback information for the expansion process. Reusing such a social knowledge aims at expanding user queries with their own vocabularies instead of using a public thesaurus, which is made by people who are not aware of the individual users needs and expectations.
2. We propose a Personalized Social Query Expansion framework called PSQE. This latter provides a user-dependent query expansion based on social knowledge, i.e., for the same query of two different users, PSQE will provide two different expanded queries, which will be processed by a search engine.
3. Using an evaluation on real data gathered from three different large bookmarking systems, we demonstrate the effectiveness of our framework for socially driven query expansion compared to many state of the art approaches.

The rest of this paper is organized as follows: in Sect. 2 we introduce all the concepts that we use throughout this paper. Section 3 introduces our method of query expansion using folksonomy. In Sect. 4, we discuss the different experiments that evaluate the performance of our approach. Related work is discussed in Sect. 5. Finally, we conclude and provide some future directions in Sect. 6.

2   Background and Notations

In this section, we formally define the basic concepts that we use throughout this paper namely, a bookmarks, a folksonomy, and a user profile. We also provide a formal definition of the problem we are intending to solve.

---

[1] This is an extended and revised version of a preliminary conference report that was presented in [12].

2.1   Background

Social bookmarking websites are based on the techniques of *social tagging or collaborative tagging*. The principle behind social bookmarking platforms is to provide the user with a means to annotate resources on the Web, e.g., URIs in *delicious*[2], videos in *youtube*[3], images in *flickr*[4], or academic papers in *CiteU-Like*[5]. These annotations (also called tags) can be shared with others. This unstructured (or better, free structured) approach to classification with users assigning their own labels is variously referred to as a *folksonomy* [21,28]. A folksonomy is based on the notion of bookmark, which is formally defined as follows:

Definition 1 (Bookmark). Let U, T, R be respectively the set of Users, Tags and Resources. A *bookmark* is a triplet (u,t,r) such as $u \in U$, $t \in T$, $r \in R$, which represents a user u who used a tag t to annotate a resource r.

Then, a group of bookmarks which forms a folksonomy is formally defined as follows:

Definition 2 (Folksonomy). Let U, T, R be respectively the set of Users, Tags and Resources. A folksonomy $\mathbb{F}(U,T,R)$ is a subset of the cartesian product $U \times T \times R$ such that each triple $(u,t,r) \in \mathbb{F}$ is a *bookmark*.

A folksonomy can then be naturally represented by a tripartite-graph where each ternary edge represents a bookmark. In particular, the graph representation of the folksonomy $\mathbb{F}$ is defined as a tripartite graph $\mathcal{G}(V,E)$ where $V = U \cup T \cup R$ and $E = \{(u,t,r)|(u,t,r) \in \mathbb{F}\}$. Figure 1 shows seven bookmarks provided by two users on three resources using three tags.

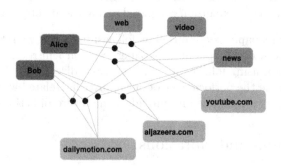

Fig. 1. Example of a folksonomy. The triples (u,t,r) are represented as ternary-edges connecting users, resources and tags.

---

[2] http://www.delicious.com/.
[3] http://www.youtube.com/.
[4] http://www.flickr.com/.
[5] http://www.citeulike.org/.

Folksonomies have proven to be a valuable knowledge for user profiling [17, 35, 41, 43]. Especially, because users tag interesting and relevant information to them with keywords that may be a good summary of their interest. Hence, in this paper, and in the context of folksonomies, the profile includes all the terms used as tags along with their weights to capture user's tagging activities. It is formally defined as follows:

Definition 3 (User Profile). Let U, T, R be respectively the set of Users, Tags and Resources of a folksonomy $\mathbb{F}(U, T, R)$. A profile p_u assigned to a user $u \in U$, is modeled as a weighted vector $\mathbf{p_u}$ of m dimensions, where each dimension represents a tag the user employed in his tagging actions. More formally, $\mathbf{p_u} = \{w_{t_1}, w_{t_2}, ..., w_{t_m}\}$ such that w_{t_m} is the weight of t_m, such as $t_m \in T \wedge (\exists r \in R \mid (u, t_m, r) \in \mathbb{F})$.

Thus, the profile includes the most relevant terms for the user and not all his activities, i.e., the documents that he has tagged. A value is associated to each term of the profile expressing its strength and importance for the given user.

Later in Sect. 3.1.4, we propose a method to assign weights to each term in the user profile in order to better define his interests.

2.2 Problem Definition

As mentioned before, query expansion consists of enriching the initial query with additional information. This expansion is generally expected to provide better search results. However, providing merely a uniform expansion to all users is, from our point of view, not really suitable nor efficient since relevance of documents is relative for each user. Thus, a simple and uniform query expansion is not enough to provide satisfactory search results for each user. Hence, having a folksonomy $\mathbb{F}(U, T, R)$, the problem we are addressing can be formalized as follows:

For a given user $u \in U$ who issued a query $q = \{t_1, t_2, ..., t_n\}$, how to provide for each term $t_i \in q$ a ranked list of related terms $\mathcal{L} = \{t_{i1}, t_{i2}, ..., t_{ik}\}$, such that when expanding the term t_i with the top k of $\mathcal{L}$, the most relevant documents are put earlier in the ranking?

3 Social Query Expansion Approach

The approach we are proposing aims at expanding user's queries in a personalized way. It can be decomposed into two parts: (i) *an offline* and (ii) *an online* part. The offline part performs the heavy computation which consists of transforming the whole social graph of a folksonomy $\mathbb{F}$ into a graph of tags where two tags are related if they are semantically related. This part is also responsible for the construction and the update of the users' profiles, for serving the online part. The online part of the approach is responsible for computing the concrete expansion using the graph of tags and the user' profiles constructed in the offline part. In the following, we describe in more details each part and we explicitly highlight our contributions.

3.1   Offline Part

The offline part is also decomposed into two facets: (i) the transformation of the social graph of a folksonomy $\mathbb{F}$ into a graph of tags, representing similarities between tags that either occur on the same resources or are shared by the same users, and (ii) the computation of the users' profiles to highlight their interests for personalizing their queries.

The approach is based on the creation and the maintenance of a graph of tags that represents all the similarities that exist between the tags of $\mathbb{F}$. There exist two kinds of approaches that propose to achieve that: (i) an approach based on the co-occurrence of tags over resources, and (ii) an approach based on their co-occurrence over users.

3.1.1   Extracting Semantics from Resources

In the first category of approaches, [24, 30, 33] state that semantically related tags are expected to occur over the same resources. For example, tags that most occur for *google.com* on *delicious* are: *search, google, engine, web, internet.*

Thus, extracting semantically related tags can be carried out by computing similarities. There exist many similarity measures [30], but all of them need pre-processing that consists of reducing the dimensionality of the tripartite graph $\mathbb{F}$ into a bipartite graph. This reduction is generally performed through aggregation methods. From the study of existing aggregation methods proposed in [30], we have chosen the *projectional aggregation* along with the *Jaccard*, the *Dice*, and the *Overlap* similarity measures to compute the similarity between tags. We choose this aggregation method because its simplicity, and it is one which gives better results in semantic information extraction [30]. Hence, we follow the same process as [30] to extract a graph of related tags from $\mathbb{F}$ according to their co-occurrence over resources:

1. Using a function $\mathcal{F}$ on the whole *folksonomy* $\mathbb{F}$ performs a *projectional aggregation* over the user dimension, resulting in a bipartite graph Tag-Resource.
2. Then, using a function $\mathcal{G}$ on the resulting bipartite graph Tag-Resource provides a graph of tags $\mathcal{T}_R$, in which each link is weighted with the similarity between tags according using the Jaccard, the Dice or the Overlap metrics [30].

Therefore, we may obtain either a graph of tags $\mathcal{T}_R$ using the *Jaccard*, the *Dice*, or the *Overlap*. Note that we do not merge the similarity measures in a same graph of tags, meaning that a graph of tags is constructed using only one similarity measure.

We end-up with an undirected weighted graph in which nodes represent tags, and an edge between two tags represents the fact that these tags occur together at least on one resource. The weights associated to edges are computed from similarities between tags as explained beforehand. This first step is illustrated in the left upper part of Fig. 2.

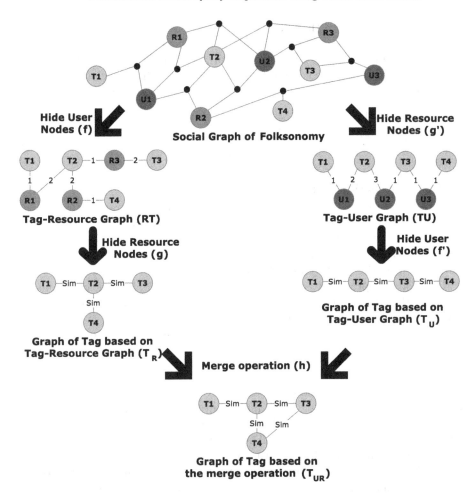

Fig. 2. Summary of the graph reduction process, which transform the whole folksonomy $\mathbb{F}$ into a graph of tags $\mathcal{T}_{\mathcal{UR}}$. The similarity values on the Figure are computed using the Jaccard measure on both graphs $\mathcal{T}_R$ and $\mathcal{T}_{\mathcal{U}}$, and using $\alpha = 0.5$ on the graph $\mathcal{T}_{\mathcal{UR}}$.

3.1.2 Extracting Semantics from Users

In the second category, [4,33] state that correlated tags are also used by the same users to annotate resources. For example, the tags *Collaborative* and *Blog* have been used 13,557 times together by users in our *delicious* dataset.

This observation is more expected to happen in certain folksonomies, where users are encouraged to upload their personal resources which leads to generate private bookmarks, e.g., a folksonomy such as *CiteULike*, *Flickr*, or *YouTube* where users are expected to upload respectively their research papers, images, and videos. Therefore, similarly to the previous approach, [33] proposes to extract semantically related tags using the following process:

1. Using a function $\mathcal{G}'$ on the folksonomy $\mathbb{F}$ performs a *projectional aggregation* over the resource dimension for obtaining a bipartite graph Tag-User.
2. Then the function $\mathcal{F}'$ is used to get another graph of tags $\mathcal{T}_\mathcal{U}$ where similarities between tags are computed using one of the three previous similarity measures.

This process is illustrated in the right upper part of Fig. 2. Notice that the structure of the graph of tags $\mathcal{T}_R$ is different from the one of the graph of tags $\mathcal{T}_\mathcal{U}$.

3.1.3   Construction of the Graph of Tag Similarities

Using only one of the two previous methods to construct a graph representing similarities between tags leads to a loss of information on one side or the other. For example, if we choose to extract related tags according to their co-occurrence over resources, we neglect the fact that there are some tags which are expected to be shared by the same users and vice versa.

Therefore, we propose to use a function $\mathcal{M}$ which is applied on the graphs of tags $\mathcal{T}_R$ and $\mathcal{T}_\mathcal{U}$ to merge them and to get a unique graph of tags $\mathcal{T}_{\mathcal{U}R}$ where the new similarity values are computed by merging the values using the *Weighted Borda Fuse (WBF)* [18]. This merge is summarized in Eq. 1, where $0 \le \alpha \le 1$:

$$Sim_{\mathcal{T}_{\mathcal{U}R}}(t_i, t_j) = \alpha \times Sim_{\mathcal{T}_R}(t_i, t_j) + (1 - \alpha) \times Sim_{\mathcal{T}_\mathcal{U}}(t_i, t_j) \qquad (1)$$

Where, $Sim_{\mathcal{T}_{\mathcal{U}R}}(t_i, t_j)$ calculates the similarity between two tags relying on the two other types of nodes, i.e., users and resources. The parameter α represents the importance one wants to give to the two types of graphs, i.e., resources or users, in the consideration of the similarity calculation. In fact, depending on the context, when computing the similarity between two tags, one may want to give a higher importance to users sharing these two tags than documents having these tags as a common tags. Another user may want to give more importance to their co-occurrence over resources than to the users sharing these tags. Depending on the nature of the folksonomy, we set α to its optimal value in order to maximize the tags semantics extraction. Finally, it should be noted that the merge is performed between graphs generated with the same similarity measure.

This step of the offline part extracts semantics from the whole social graph of $\mathbb{F}$ without a loss of information, i.e., by exploiting the co-occurrences of tags over resources and users. This step leads to the creation of a graph of tags, where edges represent semantic relations between tags. This graph will be further used to extract terms that are semantically related to a given term of a query to perform the query expansion. The contribution at this stage is the combination of the graphs resulting from resources and users to construct a better graph of tag similarities without loss of information. This is different from the existing approaches where only one graph is used.

In the following, we introduce our method of constructing and weighting the user profiles in order to personalize the expansions.

3.1.4 Construction of the User Profile

To achieve a personalized expansion, we also propose to build a user profile that consists of capturing information regarding real user interests. There are different ways to build user profiles [23,40,41]. For example, a person may be modeled as a vector of attributes of his online personal profiles including the name, affiliation, and interests. Such simple factual data provides an inadequate description of the individual, as they are often incomplete, mostly subjective and do not reflect dynamic changes [23].

Since we focus on folksonomies, the user feedback is expected to be mostly explicit (because of the tagging action, where the user explicitly assigns tags to resources).

Thus, in a folksonomy, users are expected to tag and annotate resources that are interesting to them using tags that summarize their understanding of resources. In other words, these tags are in turn expected to be a good summary of the user's topics of interests as also discussed in [2,17,23,35,37,43]. Hence, each user can be modeled as a set of tags and their weights.

The definition of a user profile is given in Definition 3. The main challenge here is *how to define the weight of each tag in the user profile?* We propose to use an adaptation of the well known *tf-idf* measure to estimate this weight. Hence, we define the weight w_{t_i} of the term t_i in the user profile as the *user term frequency, inverse user frequency (utf-iuf)*, which is computed as follows:

$$utf - iuf_{t_i,u_j} = \frac{n_{t_i,u_j}}{\sum\limits_{t_k \in \mathbf{P_u^m}} n_{t_k,u_j}} \times log\left(\frac{|U|}{|U_{t_i}|}\right) \qquad (2)$$

where n_{t_i,u_j} is the number of time the user u_j used the tag t_i.

A high value of *utf-iuf* is reached by a high user term frequency and a low user frequency of the term in the whole set of users. Note that we perform a stemming on tags before computing the profiles, to eliminate the differences between terms having the same root to better estimate the weight of each term.

User profiles are created offline and maintained incrementally. This is motivated by the fact that profiles and tagging actions are not evolving as quickly as query formulation on the system. As an analogy, it is well known that 90% of users in the social Web consume the content (i.e., query formulation), 9% update content, and 1% generate new content (profile updates) [34]. Thus, we have decided to handle the profile construction as an offline task while providing a maintenance process for keeping it up to date.

In summary, at the end of the offline part, we build two assets: (i) a graph of tags similarities which is used to represent semantically relatedness of terms, and (ii) user profiles which are leveraged in the personalization step.

3.2 Online Part

The online part of the approach is responsible for computing the concrete expansion using the graph $\mathcal{T}_{\mathcal{UR}}$ and the profiles constructed in the offline part. Before presenting our algorithm of query expansion, we propose a method to compute, on the fly, the interest of a user to a given tag.

3.2.1   Interest Measure to Tag

Having computed the similarity graph between tags and built users' profiles containing the degree to which a set of tags are representative of a user, it becomes possible to compute a degree of interest a user may have to other tags, e.g., query tags. This is useful in our approach to compute, in real time, the suitable expansions of a tag w.r.t. a given user. In our approach, this interest is seen as a similarity between the user profile $\mathbf{p_u}$ and a tag t_i. Intuitively, the computed similarity captures the interest of the user u in the query term t_i denoted $\mathcal{I}_{t_i}^u$:

$$\mathcal{I}_u(t_i) = \sum_{t_j \in \mathbf{P_u}} (Sim_{\mathcal{T}_{U\mathcal{R}}}(t_i, t_j) \times w_j) \tag{3}$$

where $Sim(t_i, t_j)$ is the similarity between the term t_i and t_j, the jth term of the user profile, and w_j is the weight of the term t_j in the profile computed during the previous process. Notice that any similarity measure can be used for computing $Sim(t_i, t_j)$, as discussed in [30]. In this work, we consider the *Jaccard*, the *Overlap*, and the *Dice* similarity measures, as discussed in the previous sections.

3.2.2   Effective Query Expansion

In this step of query expansion, we consider that the similarity between two terms t_i (a query term) and t_j (a potential candidate for the expansion of t_i), to be influenced by two main features: (i) the semantic similarity between t_i and t_j (the semantic strength between the two terms), and (ii) the extent to which the tag t_j is likely to be interesting to the considered user.

Once these two similarities are computed, a merge operation is necessary to obtain a final ranking value that indicates the similarity of t_j with t_i w.r.t. the user u. For this, several aggregation methods and algorithms exist. We choose the *Weighted Borda Fuse (WBF)* as summarized in Eq. 4, where $0 \leq \gamma \leq 1$ is a parameter that controls the strength of the semantic and social parts of our approach. Using Eq. 4, we can rank a list of terms $\mathcal{L}$, which are semantically related to a given term t_i from a user perspective.

$$Rank_t^u(t_j) = \overbrace{\gamma \times Sim_{\mathcal{T}_{U\mathcal{R}}}(t, t_j)}^{\text{Semantic Part}} + \underbrace{(1 - \gamma) \times \mathcal{I}_{t_j}^u}_{\text{Social Part}} \tag{4}$$

The effective social query expansion is summarized in Algorithm 1. Hence, for a query $q = t_1 \wedge t_2 \wedge \ldots \wedge t_m$ issued by a user u, we first get the user's profile, which is computed as explained above (Sect. 3.1.4 and Line 1 in Algorithm 1). At this stage, the purpose is to enrich each term t_i of q with related terms (line 2). Then, the objective is to get all the neighboring tags t_j of t_i in the tag graph $\mathcal{T}_{U\mathcal{R}}$ (line 3). After that (in line 4), we compute for each t_j, the ranking value that indicates its similarity with t_i w.r.t. the user u using formula 4 (line 5). Next, the neighbor list has to be sorted according to the computed values and we keep only the k top tags (line 7). Finally, t_i and its remaining neighbors must be linked with the OR ($\vee$) logical connector (line 8) and updated in q'.

Algorithm 1. Effective Social Query Expansion

Require: A folksonomy $\mathbb{F}$
      u : a User. $q = \{t_1, t_2, ..., t_n\}$: a Query.

1: $p_u[m] \leftarrow$ extract profile of u from $\mathbb{F}$
2: **for all** $t_i \in q$ **do**
3:     $\mathcal{L} \leftarrow$ list of neighbor of t_i in tag graph $\mathcal{T}_{\mathcal{U R}}$
4:     **for all** $t_j \in l$ **do**
5:       $t_j.Value \leftarrow$ Compute the ranking score $Rank_{t_i}^u(t_j)$
6:     **end for**
7:     Sort $\mathcal{L}$ according to $t_j.Value$ and keep only the top k terms in $\mathcal{L}$
8:     Make a logical OR ($\vee$) connection between t_i and all terms of $\mathcal{L}$
9:     Set the weight of the new terms t_j as the $t_j.Value$ or the TF-IDF value, depending on the choosed strategy (See Section 3.2.3)
10:     Insert $\mathcal{L}$ in q'
11: **end for**
12: **return** q'

Example 1. If a user issues a query $q = t_1 \wedge t_2 \wedge ... \wedge t_m$, it will be expanded to $q' = \{(t_1 \vee t_{11} \vee ... \vee t_{1l}) \wedge (t_2 \vee t_{21} \vee ... \vee t_{2k}) \wedge ... \wedge (t_m \vee t_{m1} \vee ... \vee t_{mr})\}$, where t_{ij} is a term that is semantically related to $t_i \in q$ and socially to u.

It should be noted that in this paper, we consider that the selection of each query term is determined independently, without considering latent term relations. Most past work on modeling term dependencies has analyzed three different underlying dependency assumptions: full independence, sequential dependence [39], and full dependence [32]. Taking into account terms dependency is part of our future works.

3.2.3 Terms Weighting

Term weighting in query expansion is challenging since there is no formal method for assigning weights to new terms. Indeed, appropriately weighting terms should result in better retrieval performance. Thus, we experiment the following two strategies for weighting new terms:

– Using the ranking values of Formula 4 as the weight of the new expanded terms. This strategy provides personalized term weight assignment while considering both semantic strength and user interest.

– Using the *Term Frequency-Inverse Document Frequency (TF-IDF)* [1] as the weight of the new expanded terms as follows:

$$tf - idf_{t_i,q} = tf_{t_i} \times log\left(\frac{|D|}{|D_{t_i}|}\right) \tag{5}$$

where tf_{t_i} denotes the term frequency of t_i in the query q. This strategy provides a uniform term weight to the query while keeping the personalizing aspect in choosing terms. Notice that weights are assigned to terms in the line 9 of Algorithm 1.

4   Evaluations

In this section, we describe the two types of evaluations we performed on our approach: (i) an estimation of the parameters of our approach to provide insights regarding their potential impact on the system, and (ii) a comparison study, where our approach is compared to the closest state of the art approaches to provide insights about the obtained results and position the proposal.

4.1   Datasets

A number of social bookmarking systems exist [21]. We have selected three datasets to perform an offline evaluation: *delicious, flickr and CiteULike*. These datasets are available and public. The interest of using such data instead of crawled data is to work on widely accepted data sets, reduce the risk of noise, and an ability to reproduce the evaluations by others as well as the ability to compare our approach to other approaches on "standardized datasets". Hereafter is the description of the different datasets.

– **Delicious:** a social bookmarking web service for storing, sharing, and discovering web bookmarks. We have used a dataset which is described and analyzed in [42][6].
– **Flickr:** an image hosting, tagging and sharing website. The *Flickr* dataset is the one used and studied in [38][7].
– **CiteULike:** an online bookmarking service that allows users to bookmark academic articles. This dataset is the one provided by the *CiteULike* website[8].

Before the experiments, we performed three data preprocessing tasks: (1) Several annotations are too personal or meaningless, such as "toread", "Imported IE Fa-vorites", "system:imported", etc. We remove some of them manually. (2) Although the annotations from delicious are easy for users to read and understand, they are not designed for machine use. For example, some users may concatenate several words to form an annotation such as "java.programming"

---

[6] http://data.dai-labor.de/corpus/delicious/.
[7] http://www.tagora-project.eu/data/#flickrphotos.
[8] http://static.citeulike.org/data/2007-05-30.bz2.

or "java/programming". We split this kind of annotations before using them in the experiments. (3) The list of terms undergoes a stemming by means of the Porter's algorithm [36] in such a way to eliminate the differences between terms having the same root. In the same time, the system records the relations between stemmed terms and original terms. As for the *delicious* dataset, we add two other data preprocessing tasks: (i) we downloaded all the available web pages while removing those which are no longer available, and (ii) we removed all the non-english web pages. This operation was performed using *Apache Tika* toolkit. Table 1 gives a description of these datasets.

Table 1. Corpus details

	Bookmarks	Users	Resources	Tags
Delicious	9,675,294	318,769	425,183	1,321,039
Flickr	22,140,211	112,033	327,188	912,102
CiteULike	16,164,802	107,066	3,508,847	712,912

4.2   Evaluation Methodology

Making evaluations for personalized search is a challenge in itself since relevance judgements can only be assessed by end-users themselves [17]. This is difficult to achieve at a large scale. Different contributions [5, 8, 25, 31] state that the tagging behavior of a user of folksonomies closely reflects his behavior of search on the Web. In other words, if a user u tags a resource r with a tag t, he will choose to access the resource r if it appears in the result obtained by submitting t as a query to the search engine. Thus, we can easily state that any bookmark (u, t, r) can be used as a test query for evaluations. The main idea of the experiments is based on the following assumption:

Proposition 1. *For a personalized query $q = \{t\}$ issued by a user u with a query term t, the relevant documents are those tagged by u with t.*

Hence, in the off-line study, for each evaluation, we randomly select $2,000$ pairs (u, t), which are considered to form a personalized query set. For each corresponding pair (u, t), we remove all the bookmarks $(u, t, r) \in \mathbb{F}, \forall r \in R$ in order to not promote the resource r in the obtained results. For each pair, the user u sends the query $q = \{t\}$ to the system. Then, the query q is enriched and transformed into q' following our approach. For the *delicious* dataset, documents that match q' are retrieved, ranked and sorted using the *Apache Lucene*. For the *Flickr* and *CiteULike* datasets, we retrieve all resources that are annotated with tags of q' while representing them according to the *Vector Space Model (VSM)*. Then, the cosine similarity is used to compute similarity between a query q' and a resource r_j.

For the *Flickr* and *CiteULike* datasets, we rank all the retrieved resources using values of the cosine similarity and we consider that relevant resources are those tagged by u using tags of q' to assess the obtained results. The random selection was carried out 10 times independently, and we report the average results.

A query expansion is expected to provide more resources as an answer to a query because of its enrichment, which generally causes an increase in the total recall. In our evaluation, we are more interested in studying the ability of the method to push relevant documents to the top of the ranking. Thus, we use the *Mean Average Precision (MAP)* and the *Mean Reciprocal Rank (MRR)*, two performance measures that take into account the ranking of relevant resources.

4.3   Study of the Parameters

We intend here to observe the parameters of our approach and estimate their optimal values. These parameters are:

- γ, which controls the semantic part and the social part in the ranking of tags for an expansion (see Eq. 4). The higher its value is, the stronger is the semantic part in tag similarity ranking, and vice versa.
- The number of tags which are suitable for the expansion.
- α, which gives either a higher importance to resources or to users, when computing the graph of tags $\mathcal{T}_{\mathcal{UR}}$. We set this parameter such that: the higher its value is, the stronger are the resources' links, and thus weaker the users links are, and vice versa (see Eq. 1).
- We evaluate two strategies for weighting the expanded terms (see Sect. 3.2.3).
- Finally, we observe the impact of the similarity measures over the search results.

We refer to our approach in Figs. 3, 4, 5, and 6 as *Personalized Social Query Expansion (PSQE)*. Also, all the Figures contain the results according to each similarity measure, and for each similarity measure, the results of the two weighting strategies are shown (this results in six curves per graph).

4.3.1   Impact of the Social Interest (γ)

The results showing the impact of the user interest w.r.t. the semantic similarity is given in Fig. 3. This latter shows the evolution of the MAP and the MRR for different values of γ, while fixing $\alpha = 0.5$ and query size to 4 for our three datasets, and using the three similarity measures. We note that the smaller the value of γ is, the better is the performance. This can be explained by the fact that the higher the value of the user interest part, the more resources that the user tags are highlighted (probably other users tag them with the same tags), and the higher is the value of the MAP and the MRR. However, we consider that neglecting the semantic part of Eq. 4 is not suitable for the following reasons: (i) First, if we fix γ to 0, we are going to neglect the semantic part, and perhaps lose the query sense (even if the potential terms to expand the query are those related to the query terms); (ii) Second, if we fix γ to 0 we are going to face cold start problems, since new users don't have an initial profile that allows us to rank terms. Thus, we choose to fix γ to 0.5 for the rest of the evaluations.

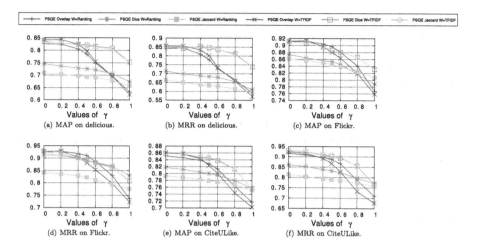

Fig. 3. Measuring the impact of the social interest (γ). For different values of γ, we fix $\alpha = 0.5$, query size $= 4$ and we use the three similarity measures and the two weighting strategies for new terms averaged over 1000 queries, using the VSM.

4.3.2 Impact of the Query Size

The objective here is to check if the length of a query impacts the obtained results. The results are illustrated in Fig. 4. Through all the experiments we have performed, it comes out that the maximum performance is achieved while adding 4 to 6 related terms to the query. Adding more than 6 related terms has no impact on the quality of the results when using values of Eq. 4 as weight

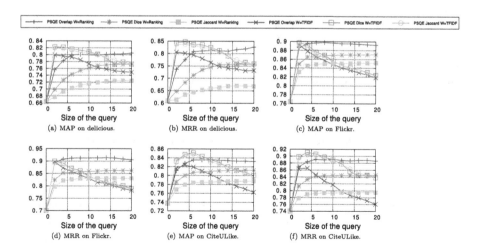

Fig. 4. Evaluating the impact of the query size on the expansion. For different values of the query size, we use $\gamma = 0.5$, $\alpha = 0.5$ and our two strategies of weighting new terms.

for new term. This has even a negative impact when using TF-IDF values for term weighting as Fig. 4 shows. For the first case, this is due to the fact that the weight of the added terms is close to 0 (we remind that the weight of the added terms is the value of Eq. 4). Hence, this makes it natural and intuitive to pick a value in the provided interval, between 4 and 6.

4.3.3   Impact of the Users and Resources (α)

The importance of users and resources on the way the expansion is performed can be tuned by the parameter α of Eq. 1. Fixing $\alpha = 0$ considers only links between tags based on common users while fixing $\alpha = 1$ considers only links between tags based on common resources. The results regarding this parameter are illustrated in Fig. 5, where the MAP and the MRR's behaviors are quite different on the three datasets.

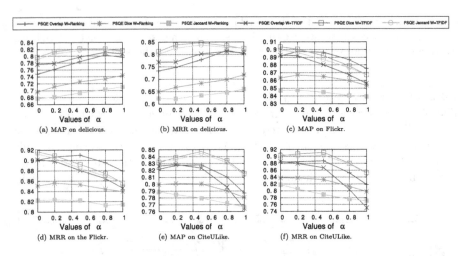

Fig. 5. Evaluating the impact of the users/resources on the expansion. For values of α, using the three similarity measures, $\gamma = 0.5$, query size $= 4$ and for our two strategies of weighting new terms.

Indeed, in the *delicious* dataset, the values of the MAP and MRR increase by increasing the value of α using both the *Jaccard* and the *Dice* similarities achieving an optimal performance at $\alpha = 1$. As for *Flickr* and *CiteULike*, the optimal performance is achieved for $\alpha = 0.2$ and $\alpha = 0.5$ respectively. We believe that this is due to the fact that in social bookmarking systems like *delicious*, users are expected to share and annotate the same resources (URLs in *delicious*) to give rise to less private resources. Therefore, annotations are expected to occur more on resources than on users. However, in social bookmarking systems like *Flickr* and *CiteULike*, users are expected to upload their own resources (images and papers) resulting in more private resources. Thus, annotations are expected to occur more on users than on resources, a property which has been also observed and reported in [16].

4.3.4   Impact of the Weight of Terms

In Sect. 3.2.3, we explain that we experiment two strategies for weighting the new expanded terms by either (i) using value of Formula 4, or (ii) the *TF-IDF* value using Formula 5. We note that the performances follow almost the same distribution while varying γ and α in Fig. 3 and 5, and for our three similarity measures over our three datasets. However, we report that each time, the *TF-IDF* weighting strategy provides better performance. Hence, we conclude that personalizing the term weighting is less advantageous and less efficient comparing to a uniform weighting approach as used in the second strategy.

4.3.5   Impact of the Similarity Measures

The behavior of the performance seem to be the same for the three similarity measures with each time a small advantage to the *Dice* measure. Hence, taking into account the ratio between all the entities to which two tags are associated together versus the union of these entities leads to a better estimation of the similarity in folksonomies.

4.4   Comparison with Existing Approaches

Our objective here is to estimate how well our approach meets the users' information needs and compare its retrieval quality to that of other approaches, objectively. Our approach is evaluated using the optimal values computed in the previous section and using our two strategies of term weighting as explained in Sect. 3.2.3. The results are illustrated in Fig. 6 as *"PSQE-W = Ranking"* for the first strategy and *"PSQE-W = TFIDF"* for the second strategy, where we select four baselines for comparison as described in the following. Note that we choose the parameters that give the optimal performance for each of these baselines.

4.4.1   PSQE vs NoQE

The first approach for comparison is that with no query expansion or personalization. Documents that match queries are retrieved, and ranked as explained above. We report the following improvements:

> ***Delicious*** **dataset:** we obtain an improvement of almost 13% of the MAP and 18% of the MRR for our first strategy of term weighting using the Overlap similarity measure, and an improvement of almost 16% of the MAP and 24% of the MRR for our second strategy of term weighting using the Dice similarity measure.
>
> ***Flickr*** **dataset:** we obtain an improvement of almost 13% of the MAP and 21% of the MRR for our first strategy of term weighting using the Overlap similarity measure, and an improvement of almost 14% of the MAP and 21%

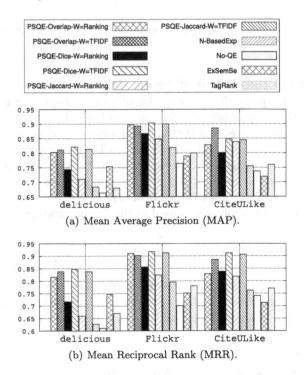

(a) Mean Average Precision (MAP).

(b) Mean Reciprocal Rank (MRR).

Fig. 6. Comparison with the different baselines of the MAP and MRR, while fixing $\gamma = 0.5$ and query size $= 4$, using the delicious, Flickr, and CiteULike datasets. We choose the optimal value of α for each similarity measure.

of the MRR for our second strategy of term weighting using the Dice similarity measure.

CiteULike **dataset:** we obtain an improvement of almost 10% of the MAP and 7% of the MRR for our first strategy of term weighting using the Jaccard similarity measure, and an improvement of almost 15% of the MAP and 14% of the MRR for our second strategy of term weighting using the Overlap similarity measure.

Thus, it is clear that the query expansion has an evident advantage compared to a strategy with no expansion. We refer to this approach as **NoQE** in Fig. 6.

4.4.2   PSQE vs N-BasedExp

The second approach is the neighborhood based approach, which is based on the co-occurrence of terms over resources. This approach consists of enriching the query q with the most related terms without considering the user profile. Thus, queries are enriched similarly for each user. Our approach significantly outperform the neighborhood based approach as follows:

Delicious **dataset:** we obtain an improvement of almost 12% of the MAP and 19% of the MRR for our first strategy of term weighting using the Overlap similarity measure, and an improvement of almost 14% of the MAP and 22% of the MRR for our second strategy of term weighting using the Dice similarity measure.

Flickr **dataset:** we obtain an improvement of almost 8% of the MAP and 12% of the MRR for our first strategy of term weighting using the Overlap similarity measure, and an improvement of almost 9% of the MAP and 12% of the MRR for our second strategy of term weighting using the Dice similarity measure.

CiteULike **dataset:** we obtain an improvement of almost 8% of the MAP and 5% of the MRR for our first strategy of term weighting using the Jaccard similarity measure, and an improvement of almost 13% of the MAP and 12% of the MRR for our second strategy of term weighting using the Overlap similarity measure.

Therefore, we conclude that our personalized query expansion efforts bring a considerable contribution according to an approach based on the most related terms. We refer to this approach as **N-BasedExp** in Fig. 6.

4.4.3   PSQE vs ExSemSe

The third approach is an approach proposed in [4], which is a strategy that uses semantic search with query expansion named *Expanded Semantic Search*. In summary, this strategy consists of adding to the query q, k possible expansion tags with the largest similarity to the original tags in order to enrich its results. For each query, the query initiator u, ranks results using BM25 and tag similarity scores. We implemented this strategy and evaluated it over our datasets. We refer to this approach as **ExSemSe** in Fig. 6. We report the following improvements:

Delicious **dataset:** we obtain an improvement of almost 5% of the MAP and 7% of the MRR for our first strategy of term weighting using the Overlap similarity measure, and an improvement of almost 7% of the MAP and 10% of the MRR for our second strategy of term weighting using the Dice similarity measure.

Flickr **dataset:** we obtain an improvement of almost 11% of the MAP and 16% of the MRR for our first strategy of term weighting using the Overlap similarity measure, and an improvement of almost 12% of the MAP and 16% of the MRR for our second strategy of term weighting using the Dice similarity measure.

CiteULike **dataset:** we obtain an improvement of almost 12% of the MAP and 10% of the MRR for our first strategy of term weighting using the Jaccard similarity measure, and an improvement of almost 17% of the MAP and 17% of the MRR for our second strategy of term weighting using the Overlap similarity measure.

4.4.4   PSQE vs TagRank

The fourth approach is an approach proposed in [6], which is an algorithm called *TagRank* that automatically determines which tags best expand a list of tags in a given query. We implemented this strategy and evaluated it over our datasets. We refer to this approach as **TagRank** in Fig. 6. We report the following improvements:

> *Delicious* **dataset:** we obtain an improvement of almost 18.10% of the MAP and $21, 79\%$ of the MRR for our first strategy of term weighting using the Overlap similarity measure, and an improvement of almost 20.83% of the MAP and 26.42% of the MRR for our second strategy of term weighting using the Dice similarity measure.
>
> *Flickr* **dataset:** we obtain an improvement of almost 12.20% of the MAP and $16, 67\%$ of the MRR for our first strategy of term weighting using the Overlap similarity measure, and an improvement of almost 12.94% of the MAP and 17.58% of the MRR for our second strategy of term weighting using the Dice similarity measure.
>
> *CiteULike* **dataset:** we obtain an improvement of almost 10.23% of the MAP and $8, 79\%$ of the MRR for our first strategy of term weighting using the Jaccard similarity measure, and an improvement of almost 16.49% of the MAP and 18.35% of the MRR for our second strategy of term weighting using the Overlap similarity measure.

In summary, the obtained results show that our approach of personalization in query expansion using social knowledge may significantly improve web search. By comparing the PSQE framework to the closest state of the art approaches, we show that it is a very competitive approach that mays provide high quality results whatever the dataset used. Finally, we notice that the better performance are obtained with the *Dice* similarity measure and using TF-IDF for term weighting over our three datasets.

5   Related Work

Current models of information retrieval are blind to the social context that surrounds information resources, e.g., the authorship and usage of information sources, and the social context of the user that issues the query, i.e., his social activities of commenting, rating and sharing resources in social platforms. Therefore, recently, the fields of Information Retrieval (IR) and Social Networks Analysis (SNA) have been bridged resulting in Social Information Retrieval (SIR) models [20]. These models are expected to extend conventional IR models to incorporate social information [11].

In this paper, we are mainly interested in how to use social information to improve classic web search, in particular the query expansion process. Hence, we cite in the following, the main works that deal with social query expansion:

Biancalana et al. [7] proposed *Nereau*, a Query expansion strategy where the co-occurrence matrix of terms in documents is enhanced with meta-data retrieved from social bookmarking services. The system can record and interpret users' behavior, in order to provide personalized search results, according to their interests in such a way that allows the selection of terms that are candidates of the expansion based on original terms inserted by the user.

Bender et al. [4] consider SIR from both the query expansion and results ranking and propose a model that deals more with ranking results than query expansion. Lioma et al. [27] provide Social-QE by considering the query expansion (QE) as a logical inference and by considering the addition of tags as an extra deduction to this process. In the same spirit, Jin et al. [24] propose a method in which the used expansion terms are selected from a large amount of social tags in folksonomy. A tag co-occurrence method for similar terms selection is used to choose good expansion terms from the candidate tags directly according to their potential impact on the retrieval effectiveness. The work in [29] proposes a unified framework to address complex queries on multi-modal "social" collections. The approach they proposed includes a query expansion strategy that incorporates both textual and social elements. Finally, Lin et al. [26] propose this to enrich the source of terms expansion initially composed of relevant feedback data with social annotations. In particular, they propose a learning term ranking approach based on this source in order to enhance and boost the IR performances. Note that in these works, there is no personalization of the expansion process.

Bertier et al. [6] propose *TagRank* algorithm, an adaptation of the celebrated *PageRank* algorithm, which automatically determines which tags best expand a list of tags in a given query. This is achieved by creating and maintaining a *TagMap* matrix, a central abstraction that captures the personalized relationships between tags, which is constructed by dynamically computing the estimation of a distance between taggers, based on cosine similarity between tags and items. From our point of view, the proposed solution is not really suitable, since it needs the creation and the maintenance of a *TagMap* matrix for each user and the execution of an algorithm for determining close users with a high complexity.

Finally, a more recent work by Zhou et al. [44] proposes first a model to construct user profiles using tags and annotations together with documents retrieved from an external corpus. The model integrates the word embeddings text representation, with topic models in two groups of pseudo-aligned documents. Based on user profiles, the authors built two query expansion techniques based on: (i) topical weights-enhanced word embeddings, and (ii) the topical relevance between the query and the terms inside a user profile.

6   Conclusion and Future Work

This paper discusses a contribution to the area of query expansion leveraging the social context of the Web. We proposed a new approach based on social personalization to transform an initial query q to another query q' enriched with close terms that are mostly used by not only a given user but also by his social relatives. Given a social graph (folksonomy), the proposed approach starts by creating and maintaining a similarity graph of tags, that represents semantic strength between tags. The steps required to generate this graph of tags is operated offline, before the system is ready to process any query. Once this graph is created, a user profile is also created offline and maintained online for each user. These structures are used to compute personalized expansions on the fly thanks to the combination of the semantic and social dimensions. We demonstrated the effectiveness of our approach by an intensive evaluation on three large public datasets crawled from delicious, Flickr, and CiteULike. We showed that the expanded queries built by our method provide more accurate results as compared to the initial queries, by increasing the MAP in a range of 10 to 16% on the three datasets. We also compared our method to three state of the art baselines, and we showed that our query expansion method allows significant improvement in MAP, with a boost in a range between 5 to 18%. Finally, the proposed approach is being integrated into a system called *LAICOS* [9,13], which can be easily plugged into existing social bookmarking platforms.

Even with the interest of the proposed method, there are still possible improvements that we can bring. We believe that our approach is complementary to some existing approaches in the area of SIR. Thus, we are convinced that a combination with social ranking functions such as those proposed in [10,17,22,35,43] can be of a great interest.

Conflict of Interest. The author(s) declare(s) that there is no conflict of interest regarding the publication of this paper.

References

1. Baeza-Yates, R.A., Ribeiro-Neto, B.: Modern Information Retrieval: The Concepts and Technology Behind Search, 2nd edn. Addison-Wesley Longman Publishing Co. Inc., Boston (2011)
2. Bao, S., Xue, G., Wu, X., Yu, Y., Fei, B., Su, Z.: Optimizing web search using social annotations. In: Proceedings of the 16th International Conference on World Wide Web, WWW 2007, pp. 501–510. ACM, New York (2007)
3. Belkin, N.J.: Some(what) grand challenges for information retrieval. SIGIR Forum **42**(1), 47–54 (2008)
4. Bender, M., et al.: Exploiting social relations for query expansion and result ranking. In: 2008 IEEE 24th International Conference on Data Engineering Workshop (2008)
5. Benz, D., Hotho, A., Jaschke, R., Krause, B., Stumme, G.: Query logs as folksonomies. Datenbank-Spektrum **10**, 15–24 (2010)
6. Bertier, M., Guerraoui, R., Leroy, V., Kermarrec, A.-M.: Toward personalized query expansion. In: Proceedings of the Second ACM EuroSys Workshop on Social Network Systems, SNS 2009, pp. 7–12. ACM, New York (2009)
7. Biancalana, C., Micarelli, A., Squarcella, C.: Nereau: a social approach to query expansion. In: Proceedings of the 10th ACM Workshop on Web Information and Data Management, WIDM 2008, pp. 95–102. ACM, New York (2008)
8. Bischoff, K., Firan, C.S., Nejdl, W., Paiu, R.: Can all tags be used for search? In: Proceedings of the 17th ACM Conference on Information and Knowledge Management, CIKM 2008, pp. 193–202. ACM, New York (2008)

9. Bouadjenek, M.R., Hacid, H., Bouzeghoub, M.: LAICOS: an open source platform for personalized social web search. In: Proceedings of the 19th ACM SIGKDD International Conference on Knowledge Discovery and Data Mining, KDD 2013, pp. 1446–1449. ACM, New York (2013)

10. Bouadjenek, M.R., Hacid, H., Bouzeghoub, M.: SoPRa: a new social personalized ranking function for improving web search. In: Proceedings of the 36th International ACM SIGIR Conference on Research and Development in Information Retrieval, SIGIR 2013, pp. 861–864. ACM, New York (2013)

11. Bouadjenek, M.R., Hacid, H., Bouzeghoub, M.: Social networks and information retrieval, how are they converging? A survey, a taxonomy and an analysis of social information retrieval approaches and platforms. Inf. Syst. **56**, 1–18 (2016)

12. Bouadjenek, M.R., Hacid, H., Bouzeghoub, M., Daigremont, J.: Personalized social query expansion using social bookmarking systems. In: Proceedings of the 34th International ACM SIGIR Conference on Research and Development in Information Retrieval, SIGIR 2011, pp. 1113–1114. ACM, New York (2011)

13. Bouadjenek, M.R., Hacid, H., Bouzeghoub, M., Vakali, A.: PerSaDoR: personalized social document representation for improving web search. Inf. Sci. **369**, 614–633 (2016)

14. Bouadjenek, M.R., Sanner, S., Ferraro, G.: A study of query reformulation for patent prior art search with partial patent applications. In: Proceedings of the 15th International Conference on Artificial Intelligence and Law, ICAIL 2015, pp. 23–32. ACM, New York (2015)

15. Bouadjenek, M.R., Verspoor, K.: Multi-field query expansion is effective for biomedical dataset retrieval. Database **2017**, bax062 (2017)

16. Carmel, D., Roitman, H., Yom-Tov, E.: Social bookmark weighting for search and recommendation. VLDB J. **19**(6), 761–775 (2010)

17. Carmel, D., et al.: Personalized social search based on the user's social network. In: Proceedings of the 18th ACM Conference on Information and Knowledge Management, CIKM 2009, pp. 1227–1236. ACM, New York (2009)

18. De, A., Diaz, E.E., Raghavan, V.V.: On fuzzy result merging for metasearch. In: 2007 IEEE International Fuzzy Systems Conference, pp. 1–6, July 2007

19. Efthimiadis, E.N.: Query expansion. In: Annual Review of Information Systems and Technology (ARIST) (1996)

20. Goh, D., Foo, S.: Social Information Retrieval Systems: Emerging Technologies and Applications for Searching the Web Effectively. Information Science Reference - Imprint of: IGI Publishing (2007)

21. Hammond, T., Hannay, T., Lund, B., Scott, J.: Social bookmarking tools: a general review. D-Lib Mag. **11**(4) (2005). http://www.citeulike.org/group/684/article/80546

22. Hotho, A., Jäschke, R., Schmitz, C., Stumme, G.: Information retrieval in folksonomies: search and ranking. In: Sure, Y., Domingue, J. (eds.) ESWC 2006. LNCS, vol. 4011, pp. 411–426. Springer, Heidelberg (2006). https://doi.org/10.1007/11762256_31

23. Hung, C.-C., Huang, Y.-C., Hsu, J.Y., Wu, D.K.: Tag-based user profiling for social media recommendation. In: Workshop on Intelligent Techniques for Web Personalization and Recommender Systems at AAAI 2008, Chicago, Illinois (2008)

24. Jin, S., Lin, H., Su, S.: Query expansion based on folksonomy tag co-occurrence analysis. In: 2009 IEEE International Conference on Granular Computing, pp. 300–305, August 2009

25. Krause, B., Hotho, A., Stumme, G.: A comparison of social bookmarking with traditional search. In: Macdonald, C., Ounis, I., Plachouras, V., Ruthven, I., White, R.W. (eds.) ECIR 2008. LNCS, vol. 4956, pp. 101–113. Springer, Heidelberg (2008). https://doi.org/10.1007/978-3-540-78646-7_12

26. Lin, Y., Lin, H., Jin, S., Ye, Z.: Social annotation in query expansion: a machine learning approach. In: Proceedings of the 34th International ACM SIGIR Conference on Research and Development in Information Retrieval, SIGIR 2011, pp. 405–414. ACM, New York (2011)

27. Lioma, C., Blanco, R., Moens, M.-F.: A logical inference approach to query expansion with social tags. In: Azzopardi, L., et al. (eds.) ICTIR 2009. LNCS, vol. 5766, pp. 358–361. Springer, Heidelberg (2009). https://doi.org/10.1007/978-3-642-04417-5_39

28. Lund, B., Hammond, T., Hannay, T., Flack, M.: Social bookmarking tools (ii): a case study - connotea. D-Lib Mag. 11(4) (2005). https://dblp.org/pers/hd/h/Hammond:Tony

29. Mantrach, A., Renders, J.-M.: A general framework for people retrieval in social media with multiple roles. In: Baeza-Yates, R., et al. (eds.) ECIR 2012. LNCS, vol. 7224, pp. 512–516. Springer, Heidelberg (2012). https://doi.org/10.1007/978-3-642-28997-2_53

30. Markines, B., Cattuto, C., Menczer, F., Benz, D., Hotho, A., Stumme, G.: Evaluating similarity measures for emergent semantics of social tagging. In: Proceedings of the 18th International Conference on World Wide Web, WWW 2009, pp. 641–650. ACM, New York (2009)

31. Mei, Q., Jiang, J., Su, H., Zhai, C.: Searching and tagging: two sides of the same coin? Technical report, University of Illinois at UrbanaChampaign (2007)

32. Metzler, D., Croft, W.B.: A Markov random field model for term dependencies. In: Proceedings of the 28th Annual International ACM SIGIR Conference on Research and Development in Information Retrieval, SIGIR 2005, pp. 472–479. ACM, New York (2005)

33. Mika, P.: Ontologies are us: a unified model of social networks and semantics. Web Semant. 5(1), 5–15 (2007)

34. Nielsen, J.: Participation inequality: Encouraging more users to contribute (2006)

35. Noll, M.G., Meinel, C.: Web search personalization via social bookmarking and tagging. In: Aberer, K., et al. (eds.) ASWC/ISWC -2007. LNCS, vol. 4825, pp. 367–380. Springer, Heidelberg (2007). https://doi.org/10.1007/978-3-540-76298-0_27

36. Porter, M.F.: An Algorithm for Suffix Stripping, pp. 313–316. Morgan Kaufmann Publishers Inc., San Francisco (1997)

37. Schenkel, R., et al.: Efficient top-k querying over social-tagging networks. In: Proceedings of the 31st Annual International ACM SIGIR Conference on Research and Development in Information Retrieval, SIGIR 2008, pp. 523–530. ACM, New York (2008)

38. Schifanella, R., Barrat, A., Cattuto, C., Markines, B., Menczer, F.: Folks in folksonomies: social link prediction from shared metadata. In: Proceedings of the Third ACM International Conference on Web Search and Data Mining, WSDM 2010, pp. 271–280. ACM, New York(2010)

39. Srikanth, M., Srihari, R.: Biterm language models for document retrieval. In: Proceedings of the 25th Annual International ACM SIGIR Conference on Research and Development in Information Retrieval, SIGIR 2002, pp. 425–426. ACM, New York (2002)

40. Stoyanovich, J., Amer-Yahia, S., Marlow, C., Yu, C.: Leveraging tagging to model user interests in del.icio.us. In: AAAI Spring Symposium: Social Information Processing, pp. 104–109 (2008)

41. Vallet, D., Cantador, I., Jose, J.M.: Personalizing web search with folksonomy-based user and document profiles. In: Gurrin, C., et al. (eds.) ECIR 2010. LNCS, vol. 5993, pp. 420–431. Springer, Heidelberg (2010). https://doi.org/10.1007/978-3-642-12275-0_37

42. Wetzker, R., Zimmermann, C., Bauckhage, C.: Analyzing social bookmarking systems: a del.icio.us cookbook. In: Proceedings of the ECAI 2008 Mining Social Data Workshop, ECAI 2008 (2008)

43. Xu, S., Bao, S., Fei, B., Su, Z., Yu, Y.: Exploring folksonomy for personalized search. In: Proceedings of the 31st Annual International ACM SIGIR Conference on Research and Development in Information Retrieval, SIGIR 2008, pp. 155–162. ACM, New York (2008)

44. Zhou, D., Wu, X., Zhao, W., Lawless, S., Liu, J.: Query expansion with enriched user profiles for personalized search utilizing folksonomy data. IEEE Trans. Knowl. Data Eng. (2017)

A Data Services Composition Approach for Continuous Query on Social Media Streams

Guiling Wang[1,2]([✉]), Xiaojiang Zuo[1], Marc Hesenius[3], Yao Xu[2], Yanbo Han[1], and Volker Gruhn[3]

[1] Beijing Key Laboratory on Integration and Analysis of Large-Scale Stream Data,
North China University of Technology,
No. 5 Jinyuanzhuang Road, Shijingshan District, Beijing 100144, China
wangguiling@ict.ac.cn
[2] Ocean Information Technology Company,
China Electronics Technology Group Corporation (CETC Ocean Corp.),
No. 11 Shuangyuan Road, Badachu Hi-Tech Park, Shijingshan District,
Beijing 100041, China
[3] paluno - The Ruhr Institute for Software Technology,
University of Duisburg-Essen, Schützenbahn 70, 45127 Essen, Germany

Abstract. We witness a rapid increase in the number of social media streams due to development of Web2.0, IoT and Cloud Computing technology. These sources include both traditional relational databases and streaming data from messaging infrastructure. We would like to use multiple social media streams to answer complex queries to enable information sharing and intelligence gathering for better collaboration. For this purpose, we adopt data services as the basic abstraction for both traditional relational databases and data streams retrieval. A flexible continuous data service model with continuous query as service operation is proposed. Service operation instance is modeled as a view defined on data streams. In the view, data part and time synchronization part are separated from each other. Based on the continuous data service model, we proposed a continuous data service composition algorithm for answering queries across data streams and relational data. The main idea is to find the contained rewriting of user query on views satisfying both data part and time synchronization part containment relationship. We also present use case and experimental studies that indicate that the approach is effective and efficient.

Keywords: Data streams · Query rewriting · Data services
Service composition · Continuous query

1 Introduction

Recent years have witnessed a social media streams boom with the increasing number of social platforms (e.g. Twitter, Facebook, Weibo, WeChat). A lot of

© Springer-Verlag GmbH Germany, part of Springer Nature 2019
A. Hameurlain et al. (Eds.): TLDKS XL, LNCS 11360, pp. 26–57, 2019.
https://doi.org/10.1007/978-3-662-58664-8_2

useful services such as crisis management, market research and location-based service are delivered by gathering, querying and analyzing these data streams. Borrowing the idea of social platforms, modern enterprises also use enterprise social platforms to enable information sharing and intelligence gathering for better collaboration.

Web services technology is a general medium for sharing data and functionality and enabling cross-organization collaboration for enterprise and web-based systems including social platforms. Data service [1] or data-providing service [2] is a kind of services that allow query-like access to organization's data. Data services provide a more flexible, controlled and standardized approach to access or query organization's data sources without exposing organization's databases directly. Furthermore, when a user require access data sources across organizations, several services can be composed to answer user query [3–5].

Social media streams have the features of data streams which are continuously arriving, rapid, time-varying, possibly unpredictable and unbounded [6]. Traditional access methods are no longer able to cope with the complexity of social streaming data. Here comes the questions of continuous query on social media streams using data services approach. How to access and share social streams using a data service approach? How to model continuous data services to provide query-like access to the underlying data streams? And how to compose continuous data services to provide social stream query across organizations? Is there a unified data service model of query and conjunctive query for both persistent relations and transient data streams?

Though there are some related work on data service modeling and composition to support data sharing, the difference between data streams and traditional data sources makes the problem of accessing and sharing social data streams challenging based on data service modeling and composition approach. (1) Different from traditional data service model, data services for queries on data streams need to continuously update service responses and consider temporal constraints. (2) In order to answer queries over multiple data sources, one feasible solution is to model services as parameterized views over data sources, and compose the services using a query rewriting approach based on the service model. Because most of the stream query languages do not support views [7], how to model data services as views over data streams is not trivial. And what's more, the composition algorithms need to be proposed to answer queries over multiple data sources automatically satisfying both data and temporal constraints.

In this paper, we propose a data service composition approach for continuous conjunctive query on social media streams. The proposed approach largely draws from experiences in the areas of data service composition, answering queries over views and views over data streams. The contributions of this paper are as follows:

1. *Continuous Data Service Model*—We introduce a continuous data service model. Service operation inputs are not modeled as fixed query conditions. They are arbitrary query conditions modeled as a set of optional attributes of the underlying data model and condition predicates. The service model is flexible because any continuous query on the underlying data streams that

can be transformed into *SyncSQL* expression can be expressed as a service operation. Thus the instance of the service operation can be modeled as a view defined on data streams. The novelty of this data service model lies in two aspects: (1) It is based on a synchronized relation model and *SyncSQL* query semantics. (2) We model the inputs and outputs of a data service as the synchronized relation stream set subscribed from or published to message queue. Service operations support time constraints and are modeled as *SyncSQL* query.

2. *Answering Query Across Data Streams*—We propose a novel continuous data service composition algorithm for answering queries across data streams. The idea behind the algorithm is the following: (1) We transform services and service instances into views on data streams. Every view has two components: data part and time synchronization part. (2) We find the contained rewriting using algorithm for "answering queries using views" on traditional relational data. Check the containment relationship between time synchronization part of the query and the rewriting. (3) We determine input and output parameters and values of service operation and add new attribute constrains to the view of a service when the service is instantiated in the algorithm.

3. *Implementation and Evaluation*—We describe an implementation, a use case and provide a performance evaluation of the proposed approach.

The rest of this paper is organized as follows: In Sect. 2, we motivate the need for conjunctive query across social media streams, discuss the underlying challenges, and overview the proposed approach. In Sect. 3 we describe our model for continuous data services. In Sect. 4, we propose a query rewriting approach and the corresponding algorithms (SBucket and SMiniCon) for processing queries over data services two algorithms. In Sect. 5, we describe our implementation and evaluate our approach. We overview related work in Sect. 6. We provide concluding remarks in Sect. 7.

2   Motivation and Overview of the Approach

In this section, we first describe a motivating scenario from social media application for ocean transportation information services we use throughout the paper. Then, we discuss the challenges to be addressed and give the overview of the approach.

2.1   Motivation

Borrowing the idea of Internet social media platforms, some modern ocean information companies develop enterprise social platforms to enable information sharing and intelligence gathering for better collaboration. Various vessels share their location and events information on the platform. Various applications collect data like vessel trajectories, vessel basic information and so on from social media platform and also their traditional

resource management systems. Among these data sources, the data stream
vesseltraj(mmsi, long, lat, speed) records trajectory points of a vessel,
where mmsi is the Maritime Mobile Service Identity, long and lat is the longitude and latitude of the vessel location, and speed is the vessel's speed. The
relation data vesselinfo(mmsi, imo, callsign, name, type, length, width,
positionType, eta, draught) records static information of ships including the
mmsi, the International Maritime Organization (imo) code, call sign, name, type,
length, width, the Estimated Time of Arrival (eta) and draught of the vessel.
The relation data vesseltravelinfo(imo, dest, source) records the destina-
tion and the identification of the position message source.

These data streams are subordinate to different management domain and
won't expose full data access interface of their data sources directly. They provide
access to the set of services with constraints described in Table 1. Note that
in this table, the value of mmsi is simplified from 9 digits to 4 digits for the
convenience of reading.

Table 1. Data services in the ocean data query scenario

Service	Functionality and constraints	Formal expression of the underlying data streams
DS$_1$	Query on those vessels whose *imo* number less than 2000 with a time-based sliding window of window size 5 s and slide size 2 s	*vesselinfo(mmsi, imo, callsign, name, type, length, width, positionType, eta, draught), vesseltraj(mmsi, long, lat, speed), imo < 2000, wsize(5), slide(2)*
DS$_2$	Query on those vessels whose *imo* is greater than 3000 with a time-based sliding window of window size 5 s and slide size 2 s	*vesseltravelinfo(imo, dest, source), imo > 3000, wsize(5), slide(2)*
DS$_3$	Query on those vessels whose speed is less than 30 km/h with a time-based sliding window of window size 5 s and slide size 1 s	*vesseltraj(mmsi, long, lat, speed), speed < 30 km/h, wsize(5), slide(1)*
DS$_4$	Query on those vessels whose *imo* number is greater than 3000 and speed is greater than 30 km/h with a time-based sliding window of window size 5 s and slide size 2 s	*vesseltravelinfo(imo, dest, source), vesselinfo(mmsi, imo, callsign, name, type, length, width, positionType, eta, draught), imo > 3000, wsize(5), slide(2)*
DS$_5$	Query on those vessels whose *mmsi* is greater than 1000 with a time-based sliding window of window size 5 s and slide size 4 s	*vesselinfo(mmsi, imo, callsign, name, type, length, width, positionType, eta, draught), vesseltraj(mmsi, long, lat, speed), mmsi > 1000, wsize(5), slide(4)*

The underlying data streams of DS$_1$ are vesselinfo and vesseltraj. They
have constraint that imo is less than 2000 with a time-based sliding window
of window size 5 s and slide size 2 s. The underlying data stream of DS$_2$ is
vesseltravelinfo. This data stream has constraints that the imo is greater
than 3000 with window size 5 s and slide size 2 s. The underlying data stream of
DS$_3$ is vesseltraj. This data stream has constraints that the speed is less than
30 km/h with window size 5 s and slide size 1 s. The underlying data streams of

DS_4 are `vesseltravelinfo` and `vesselinfo`. This service has constraints that the `imo` is greater than 2000, with window size 5 s and slide size 2 s. The underlying data streams of DS_5 is `vesselinfo` and `vesseltraj`. They have constraints that the `mmsi` is greater than 1000 with window size 5 s and slide size 4 s.

Those services with sliding window constraints continuously push output to the service consumer once the consumer create a connection with the service producer. The output is the query results in range of the configured window size and will be updated every slide size. So we call these services "continuous data services".

Now assume the following query asks for vessels which have outstanding speed over a sliding window. Note we express the query as conjunctive queries extended with time-based sliding-window semantics. The join predicates in this notation are expressed by multiple occurrences of the same variables.

Q(mmsi, draught, dest, speed):-vesselinfo(mmsi, imo, callsign, name,

type, length, width, positionType, eta, draught), vesseltraj(mmsi, long,

lat, speed), vesseltravelinfo(imo, dest, source), speed $\geq$ 40,

wsize(5), slide(4)

Because the data sources can't be accessed directly, we can't join the data sources directly using the existing data stream management systems. We will discover the related services to answer query. Obviously service DS_3 is not useful to satisfy this query request, because DS_3 has information only on vessels whose speed is less than 30 km/h whereas we are interested in vessels which has speed greater than 40 km/h. Although DS_1 is relevant to user query, it only has `mmsi`, `draught`, `speed` information and needs to retrieve `dest` information by invoking other service like DS_2 and DS_4. However, DS_1 only has information on vessels with `imo` less than 2000, while DS_2 and DS_4 have information on vessels with `imo` greater than 3000, meaning DS_1 and DS_2, DS_4 are disjoint. So service DS_1 is also not useful to answer this user query. We are left with two possible plans to use the services to answer this query. Firstly invoke DS_5 to retrieve the list of vessels with a sliding-window of window size 5 s and slide size 4 s. Then invoke DS_4 where `imo` is greater than 3000 with a sliding-window of window size 5 s and slide size 2 s. Results from both services are joint to answer Q. Note that the sliding-window constraints of DS_4 and DS_5 are different, we also need to judge if the joint results can satisfy the query requirement. In the same way, we will also find that DS_5 and DS_2 can be joint to answer Q.

2.2 Overview of the Approach

Some of the challenges involved in providing the above-mentioned tasks are: (1) developing a model for continuous data services. Different from traditional services, data services or data-providing services are concerned with data query and retrieval. They generate the appropriate outputs given specific inputs as query conditions. They do not provide any functionality beyond data query and retrieval, and have no side effects. Unlike traditional data sources, social media streams have features of temporality and dynamicity. This makes the problem of data service modelling challenging. The data services for social media streams query and retrieval should continuously generate outputs given specific inputs as query conditions. For query over multiple social media streams,data services model should support conjunctive query over data streams. Though some related work [2–4,8,9] have proposed query model and data service model for data query, retrieval and integration, they are not inapplicable for data stream query and integration. and (2) developing techniques for answering continuous queries over continuous data services. Users should be relieved from the burdensome task of selecting, composing, and invoking data services. Given a continuous query over multiple social media streams, it will be automatically executed by selecting and orchestrating the right data services.

The overview of our approach is presented in Fig. 1. Continuous data services are modeled as views over streams. Streams are represented as tagged streams which will be introduced in details in Sect. 3.1. The details of the components of a continuous data service will be introduced in Sect. 3.4. As the same as table view in database, a view over streams can be seen as a function that maps a set of input data streams into an output derived stream. In Fig. 1, this is represented as that the data service subscribes the input streams and publishes the output stream. Each service has service operations and data and/or time constraints description on the input/output streams. User queries can be transformed as *SQL*-like query over data streams with time constraints (we use Synchronized *SQL* query language *SyncSQL* [7] in this paper, a closed language to express composable queries over data streams). The mediator selects the services that can be combined to answer the posed query using the techniques of query rewriting on views, which will be introduced in Sect. 4. Then, it generates a composite service as an execution plan for the query, execute the composite service and push results to user continuously. The composite service can be deployed as a new continuous data service.

3 Model of Continuous Data Service

3.1 Data Model

We use the synchronized relation model for describing the contents of data stream sources. The data model includes:

1. S and $\Re$(S). S is a tagged stream with the format of "Tag$\langle$Attrs$\rangle$ts", where Tag can be either insert (+), update (u), or delete (-) and ts indicates the time at which the modification takes place. For example, "+$\langle$0001, 075, ...$\rangle$1" represents a tuple in Vesseltraj inserted in the stream. The tagged stream is the incremental representation format of a raw stream. Any raw stream of a data stream source can be represented as a tagged stream. Any tagged stream S has a corresponding time-varying relation $\Re$(S). The relation is continuously modified by S's tuples.

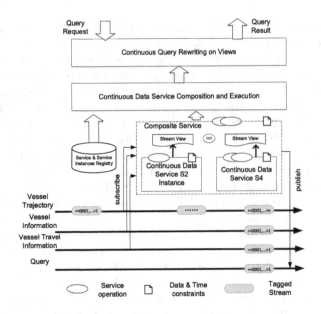

Fig. 1. Overview of the proposed approach.

2. `Attrs`. `Attrs` are the attributes of the time-varying relation $\Re(S)$.
3. `ts`. `ts` is the time point where the relation $\Re(S)$ is modified by the underlying S's tuples.
4. `Sync`. `Sync` synchronized stream is a special tagged stream "$+\langle\text{timepoint}\rangle\text{ts}$", where `timepoint` represents a time point which is the only attribute of `Sync`. For example, "$+\langle0\rangle0, +\langle2\rangle2$" is a synchronized stream indicates the time point sequences $[0, 2, 4,\dots]$. In the following paper, unit time sequence $[0, 1, 2,\dots]$ is denoted as `Sync`, $[0, 2, 4,\dots]$ is denoted Sync_2 and so on. Synchronized stream is a kind of tagged stream. So it also has a corresponding time-varying relation $\Re(\text{Sync})$.
5. $\Re_{\text{Sync}}(S)$. $\Re_{\text{Sync}}(S)$ is a synchronized relation of any arity. $\Re_{\text{Sync}_2}$ (`Vesseltraj`) is a synchronized relation that `Vesseltraj`'s tuples are reflected in $\Re_{\text{Sync}_2}$ (`Vesseltraj`) only at the time points that are specified by the synchronization stream Sync_2. Figure 2 illustrates a synchronized relation of $\Re_{\text{Sync}_2}$ (`Vesseltraj`). For example, at Time 1, $\Re_{\text{Sync}_2}$ (`Vesseltraj`) is empty and "$+\langle3001,\ \dots\rangle1$" is not inserted in $\Re_{\text{Sync}_2}$ (`Vesseltraj`) until Time 2.

`DataModel` of $\Re_{\text{Sync}}(S)$ can be represented as a tuple: $\langle\text{Attrs}, \text{SyncUnits}\rangle$, where `Attrs` = {`attr`} is a set of attributes, `SyncUnits` is the subscript index of the synchronization stream `Sync`. For example, The value of `SyncUnits` is 2 for Sync_2, 3 for Sync_3 and 4 for Sync_4 etc.

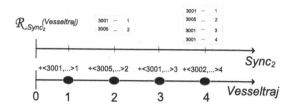

Fig. 2. A synchronized relation of $\Re_{\text{Sync}_2}$ (Vesseltraj)

3.2 Continuous Query

Continuous queries are expressed using *SyncSQL*. According to the *SyncSQL* syntax, the query Q in the motivating example can be expressed as follows:

CREATE VIEW FourUnitsSlide **AS**
SELECT 1 **AS** KeyAttr ,
      MAX(T.TimePoint) **as** currTime
FROM $\Re$(Sync$_4$) T

CREATE STREAMED VIEW R5_S4_Q **AS**
SELECT T.mmsi, T.speed , I.draught ,
      TRAVEL.dest
FROM $\Re_{\text{Sync}_4}$(Vesseltraj) T,
    $\Re$(FourUnitsSlide) N,
    Vesselinfo I,
    Vesseltravelinfo TRAVEL
WHERE T.speed $\geq$ 40 **AND**
    N.currTime - 5 < T.TS $\leq$ N.currTime

In order to represent sliding-window query Q, FourUnitsSlide is defined as a view on $\Re$(Sync$_4$) as follows: "+$\langle 1, 0 \rangle 0$, u$\langle 1, 4 \rangle 4$, u$\langle 1, 8 \rangle 8$, ...". Note this synchronized stream has only one record that is updated every 4 s.

3.3 Continuous Query Containment

Query containment and equivalence checking provide a formal framework to compare different queries in a data integration system. In relational databases, a query Q_1 is said to be contained in Q_2, denoted by $Q_1 \subseteq Q_2$, if and only if $Q_1(D) \subseteq Q_2(D)$ for any database instance D. Q_1 is *equivalent* to Q_2 if and only if $Q_1 \subseteq Q_2$ and $Q_2 \supseteq Q_1$. *Containment mapping* is used to test query containment [10]. Assume Q_1 and Q_2 are two conjunctive query, a mapping ψ from the variables of Q_1 to the variables of Q_2 is a containment mapping if (1) ψ maps every subquery in Q_1 to a sub-query in Q_2, and (2) ψ maps the head of Q_1 to the head of Q_2. $Q_1 \supseteq Q_2$ if and only if there exists a containment mapping from Q_1 to Q_2.

In stream processing system, a continuous query over n tagged streams $S_1 \ldots S_n$ is semantically equivalent to a materialized view that is defined by an SQL expression over the time-varying relations, $\Re(S_1)\ldots\Re(S_n)$ [7]. The big difference between time-varying relations and traditional relations is that the time-varying relations have arbitrary refresh conditions. The solution is to isolate the synchronization streams out of the continuous query expression. Then the containment relationship is tested in two steps: (1) test data containment using traditional query containment test method, and (2) test synchronization containment.

For example, to test the containment relationship between Q and a data service DS, we first transform the queries into *SyncSQL* expression and isolate the synchronization streams. Because *SyncSQL* expression is too long and not convenient for expressing conjunctive queries, we extend the notation of *conjunctive queries* based on a very simple form of mathematical logic [10] with synchronization notation to express the *SyncSQL* query (here we call this notation as *conjunctive queries with synchronization stream*):

$$Q(\bar{X})\text{:-}\Re(R_1)(\bar{X}_1),\ldots,\Re(R_n)(\bar{X}_n),c_1,\ldots,c_m,\text{tc},\text{Sync}_i\cap\text{Sync}_j\cap\ldots$$

In the query, $\Re(R_1)(X_1),\ldots,\Re(R_n)(X_n)$ are the sub-goals of the query. The variables in $\bar{X}$ are called *head variables* or *distinguished variables*. The c_js are interpreted atoms and are of the form $X\theta Y$, where X and Y are either variables or constants, and at least one of X or Y is a variable. The operator θ is an interpreted predicate such as $=,\leq,<,\neq,>$ or $\geq$. tc is the range time constraint. Sync_i is the synchronization stream applied on $R_1\ldots$ to R_n.

Accordingly, the query in Sect. 2 is:

$$Q(\text{mmsi},\text{draught},\text{dest},\text{speed})\text{:-}\Re(T),I,\text{TRAVEL},\text{speed}\geq 40,5,\text{Sync}_4$$

and an example data service is:

$$DS(\text{mmsi},\text{speed})\text{:-}\Re(T),I,\text{mmsi}>3000,\text{speed}\geq 30,5,\text{Sync}_1$$

To test the containment of DS and Q, we first test containment of data part of DS and Q. Because any tuples satisfied by the selection and projection conditions of Q also satisfied DS, the data part of Q is contained in data part of DS. The synchronization relation part of Q ($\Re_{\text{Sync}_4}(T)$) is contained in the synchronization relation part of DS ($\Re_{\text{Sync}_1}(T)$). We can conclude that Q is contained in DS.

3.4   Continuous Data Service

We model a continuous service as a view defined on the underlying data streams. Any service subscribes one or multiple data streams or database tables, which is defined as Subs. Any service has zero to multiple operations in which inputs, outputs, window range, window slide size should be defined. Input and output parameters are from the attributes of the underlying synchronized relations corresponding with Subs. Given a specific user inputs, the service has an associated instance. A service instance can also be defined as a view on the underlying data streams, which can be expressed as a *SyncSQL* query over the underlying synchronized relations. Each service can have multiple instances and thus can express multiple *SyncSQL* queries. Every service instance publishes one tagged stream on message queue.

Such service can be expressed as follows: DS = $\langle$ID, SubS, PubS, Ops$\rangle$, where:

1. ID is the unique identity of the service.
2. SubS is the stream set of the service subscribed from message queue. SubS = $\{\langle S_{sub}, \text{DataConstrs}, \text{TimeConstr}\rangle\}$, where DS_{sub} is a tagged stream defined in Sect. 3. A Data model $\langle$Attrs, SyncUnits$\rangle$ is corresponding with a time-varying relations $\Re(S_{sub})$. DataConstrs and TimeConstr are the constraints applied on content and time of the tagged stream.
3. PubS is the stream set of the service published to message queue. PubStreams = $\{\langle S_{pub}, \text{DataConstrs}, \text{TimeConstr}\rangle\}$, where DS_{pub} is a tagged stream. It is corresponding with a time-varying relation $\Re(S_{pub})$.
4. DataConstrs = $\{$DataConstr$\}$, where DataConstr = $\langle$attr, condop, constant$\rangle$. attr is the attribute of $\Re(S_{sub})$. condop can be one of the condition operator from $>, =, <, \geq, \neq, \leq$. constant is a constant value.
5. TimeConstr = $\langle$range$\rangle$, where range is range size of the sliding window of synchronized relation. Note that tumbling window and hopping window are both a special form of the sliding window. For tumbling window, range size is equal to slide size. And for hopping window, range size is a multiple of slide size.
6. Ops = $\{\langle$inputs, outputs, range, slide$\rangle\}$ is the service operations. inputs = $\{$input$\}$ are a set of attributes of DS_{sub}, the corresponding condition operator $>, =, <, \geq, \neq, \leq$ and constants. outputs = $\{$output$\}$ are a set of output parameters of the service operation. range and slide are the time constraint of the service request. A *SyncSQL* expression can be generated from Ops.

A service description DS = $\langle$ID, SubS, PubS, Ops$\rangle$ can be transformed into a view:

$$\text{DS}(\bar{X})\text{:-}\Re(S_{sub_1}), \ldots, \Re(S_{sub_n}), c_i, \ldots, c_m, tc, \text{Sync}_i \cap \ldots \cap \text{Sync}_m$$

where $\bar{X}$ is all the attributes from all DS_{sub} elements of SubS, $\Re(S_{sub_1}), \ldots,$ $\Re(S_{sub_n})$ are the underlying time-varying relation corresponding with all the elements of SubS and $c_i, \ldots, c_m$ are the data constraints applied on them. tc is the intersection of all the window range size constraints applied on them. Sync_i is the synchronization stream applied on $\Re(S_{sub_i})$. PubS of a service is determined when a service is instantiated. When a service is instantiated, the elements of the input set Ops are determined, which will add the additional data constraints on the description as the following service instance description. We describe a service using all the attributes from all DS_{sub} as the default output set. But the elements of the output set Ops are finally determined until a service is instantiated.

A service instance description of DS = $\langle \text{ID}, \text{SubS}, \text{PubS}, \text{Ops} \rangle$ can be transformed into a view:

$$\text{DS}(\bar{X})\text{inst:-}\Re(\text{DS}_{\text{sub}_1}), \ldots, \Re(\text{DS}_{\text{sub}_n}), c_1, \ldots, c_m, c_{\text{op}_1}, \ldots, c_{\text{op}_s},$$
$$\text{tc}, \text{Sync}_1 \cap \ldots \cap \text{Sync}_m, \cap \text{Sync}_1 \cap \ldots \cap \text{Sync}_t$$

$c_{\text{op}_1}, \ldots, c_{\text{op}_s}$ are data constraints from inputs of service operations. $\text{Sync}_1 \cap \ldots \cap \text{Sync}_t$ are synchronization stream from the time constraints of service operations. tc is the intersection of all the window range size constraints applied on $\Re(\text{Sub}_i)$ and from service operations.

4 Data Services Composition for Answering Continuous Query

Services composition is a very active area of research. The basic ingredient of any composite application are the software components encapsulating functionality, data, and/or a user interface. Services is a kind of software components providing a set of operations, which allow one to programmatically interact with the encapsulated functionality and/or UI. Examples of typical service components are SOAP and RESTful Web services. These software components can be composed into a composite application by invoking their operations according to a composition plan [11]. In this section, we present the data services composition algorithm used for answering continuous query across multiple data streams and relation data.

The algorithm is shown in Algorithm 1.

Algorithm 1. Continuous Data Services Composition

Input: conjunctive query Q in two parts:
   data part Q^d of the form: $Q^d(\bar{X}):\text{-}\Re(R_1)(\bar{X}_1), \ldots, \Re(R_n)(\bar{X}_n), c_1, \ldots, c_1, \text{tc}$
   synchronization part Sync_Q of the form: $\text{Sync}_Q = \text{Sync}_1 \cap \ldots \cap \text{Sync}_j$;
   a set of services S and service instances $\mathcal{S}$inst;

Output: a composed service CompS
 1: Let S be the union of S and $\mathcal{S}$inst, each element is S'
 2: using $Q^d(\bar{X})$ and a set of conjunctive views $S'^d(\bar{Y})$ as inputs of SBucket algorithm
    in or SMiniCon algorithm in Sect. 4.1, determine the input and output parameter
    values of services, find the service composition plan W which is the largest contained
    or equivalent rewriting set of Q, each element in W is a service instance description.

 3: **if** $W \neq \emptyset$ **then**
 4:    *ExecutePlan*(W)
 5:    CompS=*GenerateCompService*(W)
 6: **else**
 7:    CompS = $\emptyset$
 8: **end if**
 9: **return** CompS

We improve the Bucket algorithm and MiniCon algorithm [12] to find largest contained or equivalent rewriting set in step 2. The improved algorithms is called SBucket (Service Bucket) algorithm and SMiniCon (Service MiniCon) algorithm, and will be introduced in Sects. 4.1 and 4.2.

The execution plan W is a conjunctive views of service instances and may has some additional data constraints conditions. In step 4, to determine the execution order of service instances, we first pop out the first head variable from W and find those service instances that has this variable as join predicate. Join these service instances results as $R_1(x_1, \ldots, x_n)$ where x_i are variables from the joint service instances. Pop out the next head variable from W and find those remaining service instances and R_1 that has this variable as join predicate. Join them as R_2. Continue the above operations until there are no service instances left.

Algorithm 2. Create buckets

Input: conjunctive query Q in two parts:

    data part Q^d of the form:

    $Q^d(\bar{X}):\text{-}\Re(R_1)(\bar{X}_1), \ldots, \Re(R_n)(\bar{X}_n), c_1, \ldots, c_n, tc$

    synchronization part $Sync_Q$ of the form: $Sync_Q = Sync_1 \cap \ldots \cap Sync_n$;

    a set of views $\mathcal{V}$ transformed from service set $\mathcal{S}$ and service instance set $\mathcal{S}inst$;

Output: list of buckets

 1: **for** $1 \leq i \leq n$ **do**

 2:    Initialize $Bucket_i$ to $\emptyset$

 3: **end for**

 4: **for** each subgoal g_i in Q **do**

 5:    **for** each $V \in \mathcal{V}$ **do**

 6:       Let V be of the form:

            $V(\bar{Y}):\text{-}\Re(S_1)(\bar{Y}_1), \ldots, \Re(S_m)(\bar{Y}_m), d_1, \ldots, d_s, tc_V, sync_1 \cap \ldots \cap sync_t$

 7:       **if** $\Re(Sync_V) \supseteq \Re(Sync_q)$ and $tc_V \geq tc$ **then**

 8:         **if** g_i is an element of subgoals set of V **then**

 9:           **if** each $x \in X_i$ is also an element of $\bar{Y}$ **then**

10:              **if** the data constraints of V satisfy the data constraints of Q **then**

11:                add V into $Bucket_i$

12:              **end if**

13:           **end if**

14:         **end if**

15:       **end if**

16:    **end for**

17: **end for**

In step 5, $CompS = \langle ID, SubS, PubS, Ops \rangle$, where SubS are union of all PubS from the service instances with the data constraints. PubS and the elements of $Ops = \{\langle inputs, outputs, range, slide \rangle\}$ are determined when the service is instantiated.

4.1 SBucket Algorithm

In order to support finding relevant continuous data services or service instances, we improve the Bucket algorithm by adding the synchronization stream containment judgement and determining the service operation inputs and outputs after the relevant services are found. The approach for answering continuous query based on our data service model has three steps.

The first step constructs for each subgoal g in the query a bucket of relevant service or service instance atoms. It is shown in Algorithm 2.

The second step considers all the possible combinations of services and service instances. Each combination should include one of the service or service instance atoms from every bucket. Generate the candidate composition plans by checking if each combination is satisfiable (if there exists no self-contradictory in the same combination). Delete those that is unsatisfied.

Algorithm 3. Check whether a candidate plan is equivalent and instantiate the services

Input: candidate services and service instances composition plan $p(\bar{Y})$;
  conjunctive query $Q(\bar{X})$;
  a set of executable equivalent services and/or service instances composition plan eqCompPlans

Output: the updated result of eqCompPlans

1: Denote the intersection of data constraints of p and Q as $D \cap C$, where D is the data constraints set of p and C is the data constraints set of Q
2: Get all of the elements exist in set $D \cap C$ that don't exist in set of data constraints of p, denoted as $A = D \cap C \setminus D$. This set is the additional data constraints that should be added on p in order to be equivalent to Q
3: **if** $Q \subseteq p$ **then**
4:   **if** there exists services (not service instance) in p **then**
5:     **for** each subgoal g of p **do**
6:       **if** g is a service **then**
7:         $A = g.\mathtt{genInstance}(\bar{Y} \cap \bar{X}, A, \mathtt{sync})$
8:       **end if**
9:     **end for**
10:    **if** $A = \emptyset$ **then**
11:      add p into eqCompPlans
12:    **end if**
13:   **else**
14:    **if** $p \subseteq Q$ **then**
15:      add p into eqCompPlans
16:    **end if**
17:   **end if**
18: **end if**

The third step searches the equivalent service composition plans or the contained service composition plans, and determine the input and output parameters of service operations. Take searching the equivalent service composition plan as the example, the basic idea is to consider each candidate composition plan p, check if p ≡ Q when there exists no service atom (in other words, all atoms are service instances) in p. If there exist services in p and Q ⊆ p, we search the additional constraints that can be applied on services when we instantiate it. The concrete steps for considering each p are shown in Algorithm 3.

In step 3 and step 14, when we judge the containment relationship between the plan and query, time synchronization containment relationship is checked first.

In step 7, we use the additional data constraints A to instantiate a service. A method genInstance(output, dataConstr, timeConstr) is called to determine the input and output parameters of the service operation. In this method, the output parameter value is taken as the output parameter value of the service operation. In step 7, we take additional data constraints in A as the input parameter values of the service operation. According to the first stage of SBucket, time constraints of every service or service instance in p all contain the time constraints of Q, so SBucket algorithm takes the time constraints of Q as the time constraints of the service operation. In this method, we update A with the unsatisfied data constraints and returned. After the loop 5, all the services in p are instantiated. If the attributes of all the additional data constraints are also the data attributes of $\Re(g_{sub})$, it means that all the additional data constraints can be applied on the services, in other words, the services can satisfy the data constraints after instantiation. Otherwise, the service composition plan is abandoned.

In step 14, if Q ⊆ p, Q ⊇ p and all atoms of p are service instances, add p into equivalent result set directly.

The above algorithm is to search the equivalent plan. To search the contained composition plan, there is a difference that SBucket algorithm only consider the plans whose data constraints haven't existed in the contained result set. First, if Q ⊇ p and all atoms of p are service instances, add p into contained result set directly. If Q is not contained in p and data constraints of Q overlap with that of p, and there exist service atoms in p, we should instantiate the services. Check whether all the additional constraints can be applied on the services when instantiating them. If they can't be applied, this means that the services can not satisfy the data constraints after instantiation, in other words, the plan is not executable.

4.2   SMiniCon Algorithm

The above algorithm adopts the idea of Bucket algorithm into the composition approach for continuous query, however, as illustrated in [12], Bucket algorithm exists some redundant computing and has performance limitations. Firstly, it misses some important interactions between view subgoals by considering each subgoal in isolation. As a result, the buckets contain irrelevant views, and hence the second step of the algorithm becomes very expensive. Second, if there exists multiple homonymy predicates in query or view, Bucket algorithm would not realize that if it uses the predicate, then it has to use the predicate multiple of the query subgoals. Realizing this would save the algorithm exploring useless combinations in the second phase. The concrete examples are illustrated in Sect. 5.2.

So, we improve the MiniCon algorithm with continuous data service model in this paper. The MiniCon algorithm begins like the Bucket algorithm, considering which views contain subgoals that correspond to subgoals in the query. However, once the algorithm finds a partial mapping from a subgoal g in the query to a subgoal g_1 in a view V, it changes perspective and looks at the variables in the query. The algorithm considers the join predicates in the query (which are specified by multiple occurrences of the same variable) and finds the minimal additional set of subgoals that need to be mapped to subgoals in V, given that g will be mapped to g_1. This set of subgoals and mapping information is called a MiniCon Description (MCD), and can be viewed as a generalization of buckets.

In order to better illustrate the difference between our SMiniCon algorithm and MiniCon algorithm against continuous data service model, we simplify the explanation of MiniCon by assuming that all the attributes of the relations with the same relation name also have the same attribute name. In other words, we don't need to consider the attribute mapping, so the corresponding form of MCD as well as some properties can be simplified:

MiniCon Description 1. *An MCD for a query* Q *over a view* V *is a tuple of the form* (V, gc) *where* gc *is a subset of subgoals in* Q *which are covered by some subgoals in* V

Based on the simplified MCD definition, the following conditions to determine which subgoals should be added to the minimal additional set of subgoals:

Property 1. *The simplified MiniCon algorithm considers an MCD for* Q *over* V *only if it satisfied the following conditions:*
C1. If attribute x *of subgoal* g *is in head of* Q, *then the corresponding attribute* x *of subgoal* g_1 *must be in the head of* V
C2. If x *is not in head of* V, *and* x *is a join predicate in* Q, *then check every subgoal which includes* x *in* Q *according to this property to expand the additional set of subgoals.*

MiniCon algorithm has two phases. The first phase is to form MCD according to Property 1. The second phase is to create the query rewriting plan by combining multiple MCDs. In this phase, the algorithm only considers such subset of MCDS:

Property 2. *The MiniCon Algorithm considers the subset* mcds *of MCDs (formed as* $mcd_1, \ldots mcd_n$) *only if it satisfies the following conditions:*
(1). $gc_{mcd_1} \cup gc_{mcd_2} \cup \ldots \cup gc_{mcd_n} = \text{subgoals}(Q)$, *in other word, the union of all* gc *in* mcds *equals the subgoals set of* Q *(2). for each* $i \neq j$, $gc_i \cap gc_j = \emptyset$.

Considering the application scenarios in Sect. 2, queries and views are usually accompanied with comparison predicates. MiniCon algorithm can also support the comparison predicates through adding the following conditions to Property 1 in the first phase of forming MCD:

Property 3. *The simplified MiniCon algorithm considers the MCD for* Q *over* V *only if it satisfies Property 1 and the following conditions.*
C3. If there exists a data constraint VDC *with an attribute* x *in subgoal* g_1 *of* V, *then determine the containment relationship between* VDC *and the constraint* QDC *in* Q *having the same attribute* x. *If* VDC *can't be covered by* QDC *but intersect with it, and* x *is in the head of* V, *then form the MCD.*

In fact, the difference between a service and the corresponding view is, new attribute constrains can be added to the view when the service is instantiated. If VDC can't be covered by QDC, it doesn't mean that it can't be covered by QDC after the service is instantiated. Because it may be covered by QDC after attribute constrains are added. So there is no need to exclude the construction of MCD according to C3.

In order to instantiate the possible services in the second phase, we add projDC as an attribute of MCD. projDC is the constraints intersection of Q and V, and the attributes in projDC are all come from the attributes of V. So projDC can be described as the projection of data range of V to data range of Q. And what's more, in order to distinguish between equivalent rewriting and contained rewriting in the next phase, we add the mt as a tag of the MCD type, 0 represents the MCD can be used to generate the equivalent rewriting, while 1 represents the MCD is going to generate the contained rewriting. So we define the SMiniCon description (abbreviated as SMCD) as a tuple form (V, gc, projDC, mt).

SMiniCon Description 1. *An SMCD for a query* Q *over a continuous data service* DS *is a tuple of the form* (V, gc, projDC, mt), *where* gc *is a subset of subgoals in* Q *which are covered by some subgoals in the view* V *transformed from data service* DS, projDC *is the constraints intersection of* Q *and* V. *When the data constraints of* V *contains the data constraints of* Q, mt *is 0 to indicate the SMCD can be used to generate the equivalent rewriting. Otherwise,* mt *is 1 to indicate that the SMCD can be used to generate the contained rewriting.*

We based on the following conditions of a SMCD to determine which subgoals should be added to the minimal additional set of subgoals:

Property 4. *The SMiniCon Algorithm considers the SMCD for* Q *over* V *only if it satisfies the following conditions.*
C1. If attribute x *of subgoal* g *is in head of* Q, *then the corresponding attribute* x *of subgoal* g_1 *must be in the head of* V
C2. If x *is not in head of* V, *and* x *is a join predicate in* Q, *then check every subgoal which includes* x *in* Q *according to this property to expand the additional set of subgoals*
C3. The synchronization part of Q *is contained in the synchronization relation part of* V.
C4. If there exists a data constraint VDC *with an attribute* x *in subgoal* g_1 *of* V, *then* VDC *and the constraint* QDC *in* Q *having the same attribute* x *are not disjoint.*

Algorithm 4. Form SMCD

Input: conjunctive query Q in two parts: data part Q^d of the form:
    $Q^d(\bar{X})$:-$\Re(R_1)(\bar{X}_1),\ldots,\Re(R_n)(\bar{X}_n)$, $c_1,\ldots,c_n$, tc synchronization part $Sync_Q$ of the form: $Sync_Q = Sync_1 \cap \ldots Sync_n$; a set of views $\mathcal{V}$ transformed from services $\mathcal{S}$ and service instances $\mathcal{S}inst$;
Output: a set of SMCD SMCDS
 1: Initialize SMCDS to $\emptyset$, joinAttrs = genJoinAttrs(Q)
 2: **for** each subgoal g_i in Q **do**
 3:    **for** each V $\in \mathcal{V}$ **do**
 4:      Let V be of the form: $V(\bar{Y})$:-$\Re(S_1)(\bar{Y}_1),\ldots,\Re(S_m)(\bar{Y}_m)$, $d_1,\ldots,d_m$, $sync_1 \cap$
       $\ldots \cap sync_m \cap sync_1 \cap \ldots \cap sync_t$
 5:      **if** $\Re(Sync_Q) \subseteq \Re(Sync_V)$ **then**
 6:        **if** there exists no SMCD that it's name is V and it's gc covers g_i in SMCDS **then**
 7:          create a null SMCD for V and SMCD.mt = 1
 8:          **if** the data constraints of V and the data constraints of Q are disjoint **then**
 9:            break
10:          **else**
11:            **if** the data constraints of V contains the data constraints of Q,set SMCD.mt as 0
12:          **end if**
13:          canGen, gc = genGc(g_i, V, Q, joinAttrs)
14:          **if** canGen is true **then**
15:            SMCD.gc = gc
16:            SMCD.projDC = the data constraints projection of V to Q
17:            add SMCD to SMCDS
18:          **end if**
19:        **end if**
20:      **end if**
21:    **end for**
22: **end for**
23: **return** SMCDS

Our SMiniCon algorithm also has two phases. The first phase is to form SMCD. The second phase is to combine SMCDs and instantiate the possible services. The first phase of SMiniCon is shown in Algorithm 4, it creates all possible SMCD for each V according to the SMCD Description 1 and Property 4. We first check the synchronization containment relationship in step 5 to avoid unnecessary calculations. Then a method genGc(subgoalOfQ, V, Q, joinAttrs) is called to generate values of gc and canGen that denotes whether the gc can be generated. This method extends the minimal additional set of subgoals that can be covered by V through parameters of a subgoal of Q and joinAttrs (the join predicates in Q). If canJoin equals true, the projDC would be determined and add the SMCD to SMCDS.

The second phase is shown in Algorithm 5. It considers each subset of SMCDS as smcds which satisfied the Property the same as Property 2. Each smcds covers all subgoals of Q, however, in order to determine whether the combination of SMCD can generate an executable rewriting plan, we need to determine whether there is an intersection among the data ranges projected to Q (i.e. projDC) by each SMCD in the combination. We take A as the intersection result of all the projDC, if there is no constraint intersection for one attribute, A will be empty. For example, if $projDC_1$ is "mmsi > 3000, speed > 20 km" and $projDC_2$ is "mmsi < 2000, speed < 40 km", there is no constraint intersection for mmsi, so the value of A is empty. In other word, the data ranges are disjoint for this combination, so the combination can't generate an executable plan.

Algorithm 5. Find Executable Plans

Input: SMCDS:
Output: executable equal plans execEqlPlans and executable contained plans execContainPlans

1: Initialize execEqlPlans and execContainPlans to $\emptyset$
2: **for** each subset smcds of SMCDs (formed as $smcd_1$, $smcd_2 \ldots smcd_n$) that the union of all gc in smcds equals the subgoals of Q and for every $i \neq j$ $gc_i \cap gc_j = \emptyset$ **do**
3:     A = $smcd_1$.projDC $\cap$ $smcd_2$.projDC $\cap \ldots \cap smcd_n$.projDC
4:     p = $\emptyset$
5:     **if** A includes all attributes in the union set of all smcd.projDC **then**
6:         A = A $\setminus$ ($smcd_1$.V.c $\cup$ $smcd_2$.V.c $\cup \ldots \cup smcd_n$.V.c)
7:         **if** there exists smcd such that smcd.V is a service **then**
8:             **for** each smcd that smcd.V is a service **do**
9:                 A = smcd.V.genInstance($\bar{Y} \cap \bar{X}$, A, sync)
10:             **end for**
11:         **end if**
12:         **if** A = $\emptyset$ **then**
13:             p = create a executable plan consists of all smcd.V in smcds
14:             **if** all smcd.mt in smcds == 0 **then**
15:                 add p to execEqlPlans
16:             **else**
17:                 add p to execContainPlans
18:             **end if**
19:         **end if**
20:     **end if**
21: **end for**
22: **return** execEqlPlans and execContainPlans

If the combination can generate an executable plan, we update A as additional data constraints by eliminating its own constraints of each smcd.V in smcds in step 6, then we use A to instantiate the possible services in the combination. The method genInstance in step 9 is the same as the genInstance in Algorithm 3. If the A is digested to be empty by the services, it can create an executable rewriting plan consisting of all views in smcds (services are instantiated). Otherwise, there will be no executable plans to generate. Finally, the algorithm uses the mt attribute to determine whether the rewriting is an equivalent rewriting or an contained rewriting. If mt are all equals 0 in the combination, the rewriting must be an equivalent rewriting, and conversely, the rewriting is a contained rewriting, note that the SMiniCon algorithm only consider contained rewriting plans whose data constraints haven't appeared yet in contained result set which is similar to the SBucket algorithm.

5   Implementation and Evaluation

In this section, we first describe an implementation of our approach. Then we provide a use case and experimental evaluation.

5.1   Implementation

The architecture of our system is shown in Fig. 3. Users interact with the system through a web based interface. The interface enables users to register and browse relational databases and data stream sources. The registered information about the data sources are managed in the service registry. Users can formulate queries either using *SyncSQL* templates or by combining such templates into conjunctive query.

When a query is posed, the query rewriter module uses the information from service registry to generate the service composition plans and determine the inputs of the services. Implementation of the data service composition algorithm is available on Github[1]. The service executor module is responsible for determining the invocation and join order of the services and service instances.

Every service is implemented as a Spark Streaming [13] job. The underlying data streams are subscribed (represented as "sub" in the Figure) by the service using Kafka [14]. And the outputs of a service are published (represented as "pub" in the Figure) to Kafka, which can be subscribed by later services. The underlying data sources (relational databases, NoSQL databases and stream data) are registered as continuous query services by publishing to Kafka and being processed in spark streaming job after subscriptions (relational database can also registered as a common data service directly, we provide CRUD operations for it). For those Web based clients, we expose continuous data service as REST-like API [15] over HTTP protocol based on a Web-based push technology - Sever-Sent Events (SSE) [16]. It allows the service to push query results to clients continuously. The client sends a request to a service and opens a single long-lived HTTP connection. The service then sends data continuously to the client without further action from the client.

5.2   Case Study

In this section, we take the example introduced in Sect. 2 as the use case to introduce how our approach works.

$$DS_1 . SubS = \{ (\Re_{Sync_2}(T), null, 5), (I, \{imo < 2000\}, 5) \}$$

---

[1] https://github.com/declouddataservice/servicecomposition.

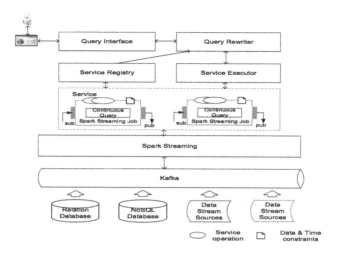

Fig. 3. Architecture of the implementation.

The *SyncSQL* expression of the query on the underlying data sources can be generated from $DS_1.SubS$ as a streamed view as follows:

CREATE STREAMED **VIEW** DS_1_Data **AS**
SELECT mmsi , imo , . . .
FROM $\Re_{Sync_2}$ (Vesseltraj) T,
　　　NowView N,
　　　Vesselinfo I
WHERE I . imo < 2000 **AND**
　　　N . currTime - 5 < T . TS ≤ N . currTime

Every instance of DS_1 can be re-written as a *SyncSQL* expression on DS_1_Data. For example, assume the outputs of Ops of an instance of DS_1 are {mmsi, draught, speed} and no input parameters. range and slide are 5 and 2 separately.

DS_1.Pubs and the *SyncSQL* expression of the query on the underlying data sources can be generated as follows:

$$DS_1.PubS = \left\{ \left(\Re_{Sync_2}\langle mmsi, draught, speed\rangle, \{imo < 2000\}, 5 \right) \right\}$$

This instance of DS_1 can be expressed as :

CREATE STREAMED **VIEW** R5_S2_DS_1inst **AS**
SELECT mmsi , draught , speed
FROM $\Re_{Sync_2}$ (Vesseltraj) T,
　　　NowView N,
　　　Vesselinfo I
WHERE I . imo < 2000 **AND**
　　　N . currTime - 5 < T . TS ≤ N . currTime

or in the form of:

$$DS_1 \text{inst}(\text{mmsi}, \text{draught}, \text{speed}) :\text{-} \Re(T), I, \text{imo} < 2000, 5, Sync_2$$

In a similar way, SubS, PubS and the instance of DS_2 can be expressed as follows:

$$DS_2.\text{SubS} = \{(\text{TRAVEL}, \text{imo} > 3000, 5)\}$$

The PubS of DS_2 for an instance is expressed as :

$$DS_2.\text{PubS} = \{(\Re_{Sync_2}\langle\text{dest}, \text{source}\rangle, \text{imo} > 3000, 5)\}$$

Then the instance of DS_2 is:

$$DS_2 \text{inst}(\text{dest}, \text{source}) :\text{-TRAVEL}, \text{imo} > 3000, 5, Sync_2$$

SubS of DS_3 is :

$$DS_3.\text{SubS} = \{(\Re_{Sync_1}(T), \{\text{speed} < 30\}, 5)\}$$

PubS of DS_3 for an instance is expressed as :

$$DS_3.\text{PubS} = \{(\Re_{Sync_1}\langle\text{mmsi}, \text{speed}\rangle, \{\text{speed} < 30\}, 5)\}$$

The instance of DS_3 can be expressed as follows :

$$DS_3 \text{inst}(\text{mmsi}, \text{speed}) :\text{-} \Re(T), \text{speed} < 30, 5, Sync_1$$

SubS of DS_4 is :

$$DS_4.\text{SubS} = \{(\text{TRAVEL}, \{\text{imo} > 3000\}, 5), (I, \{\text{imo} > 3000\}, 5)\}$$

PubS of DS_4 for an instance is expressed as :

$$DS_4.\text{PubS} = \{(\Re_{Sync_2}\langle\text{mmsi}, \text{draught}, \text{dest}\rangle, \{\text{imo} > 3000\}, 5)\}$$

The instance of DS_4 can be expressed as follows :

$$DS_4 \text{inst}(\text{mmsi}, \text{draught}, \text{dest}) :\text{-TRAVEL}, I, \text{imo} > 3000, 5, Sync_2$$

SubS of DS_5 is :

$$DS_5.\text{SubS} = \{(\Re_{Sync_4}(T), \text{mmsi} > 1000, 5), (I, \text{mmsi} > 1000, 5)\}$$

Assume there is no instance for service DS_5, so it is express as:

$$DS_5(\text{mmsi}, \text{long}, \text{lat}, \text{speed}, \text{imo}, \text{callsign}, \text{name}, \text{type},$$
$$\text{length}, \text{width}, \text{positionType}, \text{eta}, \text{draught}) :\text{-}$$
$$I, \Re(T), \text{mmsi} > 1000, 5, Sync_4$$

Query is expressed as Sect. 3.2. This query has sub-goals $\Re(T)$, I and TRAVEL. According to our SBucket algorithm in Sect. 4.1, the steps to answer user query are as follows:

In the first step the algorithm creates buckets for each sub-goal of Q. The contents of bucket for sub-goal $\Re(T)$ are: DS_1inst(mmsi, draught, speed) and DS_5(mmsi, imo, speed, long, ...). DS_3inst is not in this bucket because the interpreted predicates of the view and the query are not mutually satisfiable. The contents of bucket for sub-goal TRAVEL are: DS_2inst(dest, source) and DS_4inst(mmsi, draught, dest). The contents of bucket for sub-goal I are: DS_1inst(mmsi, draught, speed), DS_4inst(mmsi, draught, dest), and DS_5(mmsi, imo, speed, long, ...).

In the second step of the algorithm, we combine elements from the buckets. The first combination, involving the first element from each bucket, yields the rewriting

Q_1(mmsi, draught, speed, dest):-DS_1inst(mmsi, draught, speed),

DS_1inst(mmsi, draught, speed), DS_2inst(dest, source')

After we remove the first sub-goal, which is redundant, Q_1 finally consists of DS_1inst and DS_2inst. While the attribute imo does not appear in the head of DS_2inst, so we will not be able to apply the join predicate between vesseltravelinfo(imo, dest, source) and vesselinfo(mmsi, imo, callsign, name, type, length, width, positionType, eta, draught) in the query. Therefore, DS_2inst is not usable to answer query, and the combinations involving DS_2inst, for example Q_1, can't be established to answer the query.

Considering the second element in the left bucket yields the rewriting:

Q_2(mmsi, draught, speed, dest):-DS_1inst(mmsi, draught, speed),

DS_1inst(mmsi, draught, speed), DS_4inst(mmsi, draught, dest)

Remove the redundant sub-goal, Q_2 actually consists of DS_1inst and DS_4inst, while they are relevant to the query in isolation, their combination is guaranteed to be empty because they cover disjoint sets of vessel imo numbers. Similarly, combinations involving DS_1inst and DS5 at the same time should be excluded.

After eliminating the combinations above, we consider to yield the following rewriting:

Q_3(mmsi, draught, speed, dest):-DS_4inst(mmsi, draught, dest),

DS_5(mmsi, imo', speed, long', ...), DS_4inst(mmsi, draught, dest)

Then we remove the redundant sub-goal, add the predicate speed $\geq$ 40, and join with the synchronization stream. So we would obtain Q_3, which is the only contained rewriting the algorithm finds.

Then check the containment relationship between $Sync_{Q_3}$ = $Sync_2 \cap Sync_4$ and $Sync_Q$ = $Sync_4$. Apparently, $Sync_Q$ is contained in $Sync_{Q_3}$. So the time part of Q_3 is $Sync_2 \cap Sync_4$.

The output parameters of DS_5 are set to be variables from attributes of the underlying data stream which are also in the head of Q, which is mmsi, draught, speed. The inputs parameters of DS_5 are speed ≥ 40.

The main inefficiency of this algorithm is that it misses some important interactions between service subgoals by considering each subgoal in isolation. So the buckets contain irrelevant services, for example DS_2inst, and hence the second step of the algorithm becomes very expensive. In our SMiniCon algorithm proposed in Sect. 4.2, these interactions would be found in the first phase.

In the first phase, the algorithm creates an SMCD for each service. In order to express the SMCD briefly, we only provide the information about subgoals covered by each service. For DS_1inst, it can cover vesseltraj. DS_4inst covers vesselinfo and vesseltravelinfo synchronously, while DS_5 can cover vesselinfo or vesseltraj individually. Note that there are no any SMCD created for DS_3inst and DS_2inst according to SMCD creation condition.

In the second phase, the SMiniCon algorithm only focuses on combinations where the SMCDs cover mutually exclusive sets of subgoals in the query. Therefore we only need to consider the following rewritings yielded by usable combinations:

Q_4(mmsi, draught, speed, dest):-DS_1isnt(mmsi, draught, speed),

DS_4inst(mmsi, draught, dest)

Q_5(mmsi, draught, speed, dest):-DS_4inst(mmsi, draught, dest),

DS_5(mmsi, imo$'$, speed, long$'$, ...)

As we discussed in the SBucket algorithm, Q_4 will be excluded, so the final usable rewriting is Q_5. Meanwhile, the parameters determination is the same as Q_1. From the above, we can see that in the second phase SMiniCon only checks two combinations while SBucket algorithm needs to consider each element of the Cartesian product of the buckets.

5.3 Experimental Evaluation

In this section, we give an experimental evaluation of our approach. The goal of the experimental evaluation is to (1) analyze the factors that affect the performance of the service composition algorithm, and (2) analyze the factors that affect the execution performance of the continuous data services.

The service composition algorithm experiments are run on a computer with Intel(R) Core(TM) i5-2400 CPU 3.10 GHz and 8 GB memory. Experiments on the execution performance of the continuous data services are run on a cluster with the following configuration as shown in Table 2:

Table 2. Experimental Environment Configuration

Host role	CPU	Mem.	OS	Framework
Master	4×AMD Opteron 6128(2.0 GHz/8-core)	64 GB	CentOS 7.0.1406	CDH(5.11)/Spark 1.6.0
Slave1	4×AMD Opteron 6128(2.0 GHz/8-core)	64 GB	CentOS 7.0.1406	CDH(5.11)/Spark 1.6.0
Slave2	2×Intel(R)Xeon E5620(2.4 GHz/4-core)	64 GB	CentOS 7.0.1406	CDH(5.11)/Spark 1.6.0
Slave3	2×Intel(R)Xeon E5620(2.4 GHz/4-core)	72 GB	CentOS 7.0.1406	CDH(5.11)/Spark 1.6.0
Slave4	2×Intel(R)Xeon E5620(2.4 GHz/4-core)	72 GB	CentOS 7.0.1406	CDH(5.11)/Spark 1.6.0
Slave5	2×Intel(R)Xeon E5620(2.4 GHz/4-core)	72 GB	CentOS 7.0.1406	CDH(5.11)/Spark 1.6.0

In order to experimentally evaluate our approach, we select a set of queries and generate a set of continuous data services and service instances.

Here we use three representative queries:

1. Query 1: query movie IDs and directors of those movies that exceed 100 million dollars on box office returns.

   Q_1(Title, Year, Dir) : -Movie(ID, Title, Year, Genre), Revenue(ID, Amount), Director(ID, Dir), Amount > 100, wsize(5), slide(2)

2. Query 2: query those vessels whose speed exceed 40 km/h.

   Q_2(mmsi, draught, dest, speed) : -vesselinfo(mmsi, imo, callsign, name, type, length, width, positionType, eta, draught), vesseltraj (mmsi, long, lat, speed), vesseltravelinfo(imo, dest, source), speed > 40, wsize(5), slide(4)

3. Query 3: query *mmsi* and *callsign* of those vessels whose speed exceed 40 km/h.

   Q_3(mmsi, callsign) : -vesselinfo(mmsi, imo, callsign, name, type, length, width, positionType, eta, draught), vesseltraj(mmsi, long, lat, speed), speed > 40, wsize(5), slide(4)

According to 80/20 rule (also known as Pareto principle) [17], we generate data services and instances using a random method and enable the number of services and service instances related to user queries are about 20% of the total services and service instances.

For each query, we generate various number of data services and data service instances from 200, 400, ... to 1000. We here present results obtained by running each experiment ten times in Table 3.

Table 3. Query statistics for queries 1, 2 and 3 as the number of data sources is varied between 200 and 1000

Query	Services	Max bucket size	Plans enum.	Plans gen.	Time per plan (ms.)	Total time (Sec.)
1	200	29	3193	24	15.61	0.27
	400	59	25501	116	11.31	0.89
	600	89	94092	343	4.31	1.29
	800	118	212037	776	2.95	2.09
	1000	144	390105	1355	2.44	3.08
2	200	22	3000	83	7.78	0.54
	400	46	22351	515	2.51	1.21
	600	69	78757	1600	1.26	1.92
	800	88	191141	3668	0.90	3.30
	1000	110	335315	6238	0.86	5.60
3	200	28	462	64	4.63	0.27
	400	56	2000	231	2.65	0.59
	600	87	4582	509	1.78	0.89
	800	117	7763	903	1.18	1.04
	1000	149	12568	1422	0.95	1.31

Maximum bucket size is the average number of sources in the largest bucket created using SBucket algorithm in Algorithm 2. *Plans enumerated* is the average number of candidate rewriting plans (including the composition plans equivalent to the query and contained in the query) enumerated in the algorithm. Table 3 gives the average total time taken to generate all composition plans and the average time per composition.

Figure 4 plots the total and average time to generate all composition plans for each query against the number of data sources. We can observe that the average time per composition plan is within 20 ms. In this experiment, we select the number of data sources according to the experimental results of Q_2, because Q_2 took the longest time and is the most valuable for the experimental performance test. For Q_2, if the number of data sources is above 3000, the time reaches 6 min and 42 s to generate roughly the same number of composition plans. When the number of data sources is above 3500, the maximum capacity has been reached. When the number of data sources increases within a range of 200 to 1000, the average time to generate a composition plan does not increases with the growth of the data sources. It decreases and is within 20 ms.

We also compare the performance of SBucket algorithm and SMiniCon algorithm for Q_2 and Q_3. As shown in Fig. 5, SMiniCon algorithm outperforms SBucket algorithm apparently.

For Q_2 and Q_3, data services are generated from real data set. And the data services for Q_1 are generated from simulated data set. Here we do experiments on Q_2 and Q_3 to analyze the factors that affect the execution performance of the continuous data services.

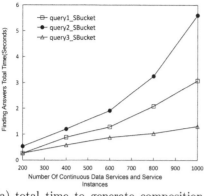
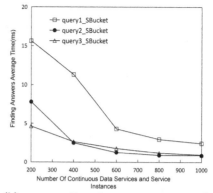

(a) total time to generate composition plans

(b) average time to generate composition plans

Fig. 4. Total and average time to generate composition plans.

We send 20, 40, 60, 80, 100 data records generated from real data sets every 10 ms to simulate data stream with the rate of 2000 (records)/s, 4000 (records)/s,..., 10000(records/s). We measure the execution time of composite service with the formula `ExecuteTime` = `WindowComputeTime-ReadSQLTime`, where `WindowComputeTime` is the time to compute the query results on the data in the current window, and `ReadSQLTime` is the time to read query request from HDFS file system or HBase. We select a composition plan randomly and run the plan ten times. The average execution time is shown in Fig. 6.

Under the situation of other conditions (window range and slide) being equal, Fig. 6 shows that the input rate of data stream have almost no impact on execution time of a composition plan. The difference between Q_2 and Q_3 is that Q_2 uses two service instances to answer the query while Q_3 only need to use one service instance. So the complexity of Q_2 is higher than the complexity of Q_3. From the experimental results in Fig. 6, we can see that the execution time of Q_2 is more than the execution time of Q_3. This indicates that the more complex the query, the lower the query performance under the same conditions.

As shown in Fig. 6, under the situation of other conditions (window slide is 4 s and input rate is 4000 (records)/s) being equal, the execution performance of the same query decreases as the window range widened. And the execution performance of the same query decreases as the window range narrowed.

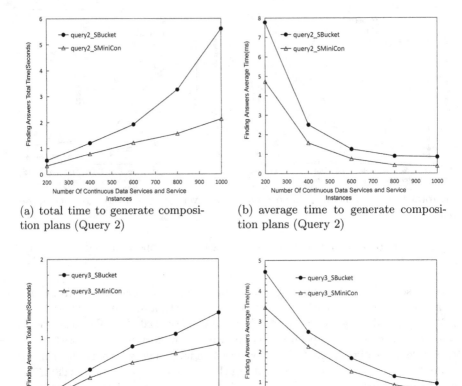

(a) total time to generate composition plans (Query 2)

(b) average time to generate composition plans (Query 2)

(c) total time to generate composition plans (Query 3)

(d) average time to generate composition plans (Query 3)

Fig. 5. Comparison of SMiniCon and SBucket.

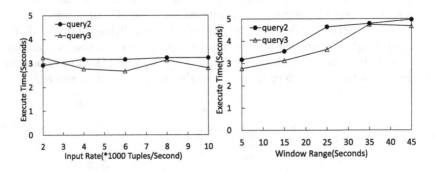

Fig. 6. Execution time of a composition plan.

6   Related Work

In this section, we compare our approach to related work in the following areas: (1) Web Service Composition, (2) Services Modeling for Data Streams, (3) Data Integration and (4) Answer Data Stream Query Using Views.

6.1   Web Service Composition

Most of the research work on web service composition focus on traditional *Effect-Providing* services or *Application-logic* services instead of *Data-Providing* services or data services. The *Application-logic* services provide business functionality such as processing an order or requesting the shipping of the goods, while the data services provide data sources with a defined data structure. There are some differences making the traditional service composition algorithms inapplicable and inefficient to data services: (1). The composition algorithm of the traditional services compose the services automatically based on their implemented functionalities, input and output constraints, preconditions, and effects (IOPEs). While for data services, they all share the same business function (i.e. data query) and have no side-effects. Hence the traditional composition algorithm can not be applied to data services directly [3]. (2). The composition constructs of *Application-logic* services are control-flow representing communications with atomic services and specify the execution order of communications. The data transformations of the inputs and outputs are embedded in the service control-flow implicitly using transformation rules. While the composition constructs of the data services are data-flow representing transfer and transformation of the output of a service as input of another service [11]. Hence the generated composition plans for application-logic services is inefficient for data services because the data-flow is often not expressed separately.

6.2   Services Modeling for Data Streams

Some work have proposed some service modeling approaches for data streams such as [18] and [19]. The data service model proposed in [18] accepts data streams as inputs and defines several stream operations such as filter, sort, merge, join and so on to process the input data streams. Our work differs from this work in the aim. The aim of the service model in [18] is for event correlation and processing. While the service model in our work is for generating composition plans automatically for data integration. [19] classifies the continuous services into four categories: producer, processor, storage and consumer. The data service model proposed in our work falls into the category of "producer". In [19], continuous services is composed as a set of continuous operations applied to a set of streams. The data to be processed are injected in the composition by "producer" services and flow through "processor" services and "storage" services until reaching the "consumer" services. The composition approach in [19] is abstracted as a task mapping problem. This is different with our work because we focus on automatically composition of "producer" services given user query request. In our previous work [20], a data stream service model is proposed to support querying and accessing data streams continuously. But the data stream service composition approach has not been discussed yet.

6.3 Data Integration

Although the existing distributed stream processing systems [13,21,22] support data stream sliding-window joins, they don't support generating the view composition plans automatically.

There are some related research work from data integration area such as Info-Master [23] and Information Manifold [24]. Our work differs with these works in many ways. Firstly, these works target toward resolving specific queries given a set of data sources, whereas in our work the focus is on constructing a composition of services that is independent of a particular input value. The composite service can be reused to answer a set of queries instead of a specific queries. Secondly, compared to previous query rewriting algorithms [12,25] that were proposed for the traditional relational data model, our composition algorithm is based on data stream model. As far as we know, our continuous data service model is the first to address the problem of composing continuous data services to support data stream integration.

Recent approaches addressed the problem of data services composition adopting the data integration approaches. There are three kinds of data integration approaches (1) Global-As-View (2) Local-As-View and (3) Global-Local-As-View (GLAV). Research work in [26] adopts Global-As-View approach. Compared to Global-As-View approach, it is much easier to add new data services for the Local-As-View approach. [27] utilized the Local-As-View approach, its composition algorithm adopted the Inverse-Rules technique to generate the composition plan. As the Inverse-Rules are computed only based on the view definition without any consideration of the query context, it has been shown typically slower than Bucket and MiniCon algorithm.

Some work are based on Local-As-View approach using the Bucket or Mini-Con techniques to generate the composition plan [2–4,8,9]. Our approach is more relevant to these research work. However, these data service models can only express snap-shot queries over data tables, continuous queries over data streams can not be supported by these models. In these work, data providing services are modeled as parameterized views over data schemas. Based on the service model, services can be composed using a query rewriting approach to answer queries over multiple data sources. Because most of the stream query languages do not support views [7], we can not model data services as views over data streams using these data services modeling methods directly. Compared with these research work on data services modeling and composition, our approach support modeling data services as views over data streams. Our approach also improved the traditional Bucket algorithm to answer continuous queries for data streams based on a set of given data services.

There are also other work use visual mashup languages or constructs as service composition approach to solve the problem of data integration [28,29].

Compared with them, the approach proposed in this paper can automatically generate composition plans given user query instead of visual programming.

6.4   Answer Data Stream Query Using Views

Some work has addressed the problem of supporting views in data stream management systems [7]. The judgement of continuous query containment relationship in our work is based on it. However, the work is limited only to solve the problem of supporting views over streams and the query plan optimization. Given a query request and a view, the algorithm in [7] determines if the query is contained in the view and rewrites the query in terms of the view to answer the query. Our work goes further. We proposes a continuous data service model which provides a flexible, controlled and standardized approach to access or query data stream. We address data stream integration problem by providing service composition approach to answer conjunctive query given a set of services. The composite service can access a set of conditions as input instead of limiting to answering specific queries.

7   Conclusion

In this paper, we presented an approach for conjunctive query on social media streams by composing continuous data services. We introduce a flexible continuous data service model with continuous query as service operation. Service operation instance is modeled as a view defined on data streams in which the data part and time synchronization part are separated from each other. A continuous data service composition algorithm is introduced for answering queries across data streams. An experimental study is provided to evaluate the scalability and performance of our approach. The results show that the algorithm scales up very well to high input rates of the underlying data streams and a large number of services and service instances. Location is one of the most important contextual features of social media streams. It is challenging to access context-based data and information over social media streams. As a future work, we plan to address location concerns when composing continuous data services, e.g. geolocating users. We also plan to consider the cost model and Quality of Service (QoS) while processing queries and composing continuous data services.

Acknowledgments. This work is supported by Beijing Natural Science Foundation No. 4172018 (Building Stream Data Services for Spatio-Temporal Pattern Discovery in Cloud Computing Environment) and National Natural Science Foundation of China No. 61672042 (Models and Methodology of Data Services Facilitating Dynamic Correlation of Big Stream Data), and University Cooperation Projects Foundation of CETC Ocean Corp.

References

1. Carey, M.J., Onose, N., Petropoulos, M.: Data services. Commun. ACM **55**(6), 86–97 (2012)
2. Vaculín, R., Chen, H., Neruda, R., Sycara, K.: Modeling and discovery of data providing services. In: 2008 IEEE International Conference on Web Services, pp. 54–61, September 2008
3. Barhamgi, M., Benslimane, D., Medjahed, B.: A query rewriting approach for web service composition. IEEE Trans. Serv. Comput. **3**(3), 206–222 (2010)
4. Zhou, L., Chen, H., Yu, T., Ma, J., Wu, Z.: Ontology-based scientific data service composition: a query rewriting-based approach. In: AAAI Spring Symposium: Semantic Scientific Knowledge Integration, pp. 116–121 (2008)
5. Zhang, F., Wang, G., Han, Y.: Automatic generation of service composition plans for correlated queries. In: 2013 10th Web Information System and Application Conference, pp. 143–149, November 2013
6. Babcock, B., Babu, S., Datar, M., Motwani, R., Widom, J.: Models and issues in data stream systems. In: Proceedings of the Twenty-First ACM SIGMOD-SIGACT-SIGART Symposium on Principles of Database Systems, pp. 1–16. ACM (2002)
7. Ghanem, T.M., Elmagarmid, A.K., Larson, P.Å., Aref, W.G.: Supporting views in data stream management systems. ACM Trans. Database Syst. **35**(1), 1–47 (2008)
8. Zhao, W., Liu, C., Chen, J.: Automatic composition of information-providing web services based on query rewriting. Sci. China Inf. Sci. **55**(11), 2428–2444 (2012)
9. Barhamgi, M., Benslimane, D., Ouksel, A.M.: Composing and optimizing data providing web services. In: Proceedings of the 17th International Conference on World Wide Web, pp. 1141–1142. ACM (2008)
10. Doan, A., Halevy, A., Ives, Z.: Principles of Data Integration, 1st edn. Morgan Kaufmann Publishers Inc., San Francisco (2012)
11. Lemos, A.L., Daniel, F., Benatallah, B.: Web service composition: a survey of techniques and tools. ACM Comput. Surv. **48**(3), 33:1–33:41 (2015)
12. Pottinger, R., Halevy, A.: MiniCon: a scalable algorithm for answering queries using views. Int. J. Very Large Data Bases **10**(2–3), 182–198 (2001)
13. Zaharia, M., Das, T., Li, H., Hunter, T., Shenker, S., Stoica, I.: Discretized streams: fault-tolerant streaming computation at scale. In: Proceedings of the Twenty-Fourth ACM Symposium on Operating Systems Principles, SOSP 2013, pp. 423–438. ACM, New York (2013)
14. Wang, G., et al.: Building a replicated logging system with Apache Kafka. Proc. VLDB Endow. **8**(12), 1654–1655 (2015)
15. Fielding, R.T.: Architectural styles and the design of network-based software architectures. Ph.D. thesis, University of California (2000)
16. Hickson, I.: Server-sent events. https://www.w3.org/TR/eventsource/. Accessed 25 Oct 2015
17. Newman, M.E.: Power laws, Pareto distributions and Zipf's law. Contemp. Phys. **46**(5), 323–351 (2005)
18. Han, Y., Liu, C., Su, S., Zhu, M., Zhang, Z., Zhang, S.: A proactive service model facilitating stream data fusion and correlation. Int. J. Web Serv. Res. (IJWSR) **14**(3), 1–16 (2017)
19. Billet, B., Issarny, V., Texier, G.: Composing continuous services in a CoAP-based IoT. In: 2017 IEEE International Conference on AI Mobile Services (AIMS), pp. 46–53, June 2017

20. Han, Y., Wang, G., Yu, J., Liu, C., Zhang, Z., Zhu, M.: A service-based approach to traffic sensor data integration and analysis to support community-wide green commute in China. IEEE Trans. Intell. Transp. Syst. **17**(9), 2648–2657 (2016)
21. Carbone, P., Katsifodimos, A., Ewen, S., Markl, V., Haridi, S., Tzoumas, K.: Apache Flink: stream and batch processing in a single engine. Bull. IEEE Comput. Soc. Tech. Comm. Data Eng. **36**(4), 28–38 (2015)
22. Confluent: KSQL: Streaming SQL for Apache Kafka. https://www.confluent.io/product/ksql/. Accessed 25 July 2018
23. Genesereth, M.R., Keller, A.M., Duschka, O.M.: Infomaster: an information integration system. SIGMOD Rec. **26**(2), 539–542 (1997)
24. Levy, A.Y., Rajaraman, A., Ordille, J.J.: The world wide web as a collection of views: query processing in the information manifold. In: VIEWS, pp. 43–55 (1996)
25. Levy, A.Y., Rajaraman, A., Ordille, J.J.: Querying heterogeneous information sources using source descriptions. In: Proceedings of the 22th International Conference on Very Large Data Bases, VLDB 1996, pp. 251–262. Morgan Kaufmann Publishers Inc., San Francisco (1996)
26. Benedikt, M., Lopez-Serrano, R., Tsamoura, E.: Biological web services: integration, optimization, and reasoning. CEUR Workshop Proceedings, pp. 21–27 (2016)
27. Thakkar, S., Ambite, J.L., Knoblock, C.A.: Composing, optimizing, and executing plans for bioinformatics web services. VLDB J. **14**(3), 330–353 (2005)
28. Wang, G., Yang, S., Han, Y.: Mashroom: end-user mashup programming using nested tables. In: Proceedings of the 18th International Conference on World Wide Web, pp. 861–870. ACM (2009)
29. Han, Y., Wang, G., Ji, G., Zhang, P.: Situational data integration with data services and nested table. Serv. Oriented Comput. Appl. **7**(2), 129–150 (2013)

DABS-Storm: A Data-Aware Approach for Elastic Stream Processing

Roland Kotto Kombi[1(✉)], Nicolas Lumineau[2], Philippe Lamarre[1],
Nicolo Rivetti[3], and Yann Busnel[4]

[1] Univ Lyon, INSA de Lyon, LIRIS UMR 5205, Villeurbanne, France
roland.kottokombi@gmail.com
[2] Univ Lyon, University Claude Bernard Lyon 1, LIRIS UMR 5205,
Villeurbanne, France
[3] Technion, Israel Institute of Technology, Haifa, Israel
[4] IMT Atlantique, Rennes, France

Abstract. In the last decade, stream processing has become a very active research domain motivated by the growing number of stream-based applications. These applications make use of continuous queries, which are processed by a stream processing engine (SPE) to generate timely results given the ephemeral input data. Variations of input data streams, in terms of both volume and distribution of values, have a large impact on computational resource requirements. DYNAMIC AND AUTOMATIC BALANCED SCALING FOR STORM (DABS-STORM) is an original solution for handling dynamic adaptation of continuous queries processing according to evolution of input stream properties, while controlling the system stability. Both fluctuations in data volume and distribution of values within data streams are handled by DABS-STORM to adjust the resources usage that best meets processing needs. To achieve this goal, the DABS-STORM holistic approach combines a proactive auto-parallelization algorithm with a latency-aware load balancing strategy.

1 Introduction

With the proliferation of connected devices (smartphones, sensors, etc.), more and more data stream sources emit real-time data with fluctuations in input rate and value distribution over time [19]. Processing these Big Data streams (volume and velocity) in soft-real time (*i.e.,* low latency), satisfying end-user performance requirements, still raises several research problems.

To process a stream set, a user can submit a query to the execution infrastructure. This query, called a *continuous query* [7,19], computes new results as new *stream elements* are generated by sources over time. Users define continuous

This work has been partially supported by the project SocioPLug (ANR-13-INFR-0003) funded by the French National Research Agency, the Association Nationale Recherche Technologie (ANRt) http://socioplug.univ-nantes.fr/index.php/SocioPlug_Project.

© Springer-Verlag GmbH Germany, part of Springer Nature 2019
A. Hameurlain et al. (Eds.): TLDKS XL, LNCS 11360, pp. 58–93, 2019.
https://doi.org/10.1007/978-3-662-58664-8_3

queries through declarative languages [2, 6, 7, 12] or, more imperatively, through a high-level language [28, 41] (Java, Python, C, etc.). In any case, these continuous queries are usually turned into direct acyclic graphs (DAG) of operators, called *workflows* or *topologies*, corresponding to execution plans [1, 2, 28].

To generate timely results, a workflow requires some resources (CPU, RAM, bandwidth). The problem is that any evolution of the input streams (in input rate or value distribution) impacts the amount of ressources needed to process it. Furthermore, end-users usually require a good end-to-end latency and no data loss, regardless of any other consideration. Since, in general, evolutions of input streams cannot be captured through a-priori knowledge, a dynamic method is required that dynamically adapts the assigned resource according to evolution of needs. Such a method has to be as precise as possible. Indeed, whatever its imprecision, its consequences are negative. On the one hand, an under-provisioning may lead to *congestion*, implying reduced throughput and increased latency, or worse, data loss [1]. On the other hand, an over-provisioning induces resource and financial wastes, while potentially generating massive network overheads [39] and resource shortage.

Industrial [9, 20], open-source [4, 5] and academic [1, 2, 6, 8, 12, 13, 36] *stream processing engines* (SPE) have been developed to simplify stream management. Nevertheless, due to a lack of holistic and automatic strategies embracing all aspects related to elasticity, most of these solutions rely on user expertise and reactivity to face critical fluctuations in input rate. In particular, this is the case for the STORM family solutions.

To adapt provisioning, three linked problems have to be considered for each operator: parallelism degree, scheduling, and load balancing. Operator parallelism defines how many threads work together to process the incoming load of one operator. Note that, until an asymptote is reached, increasing the number of threads improves system performance. The scheduling strategy assigns threads to available processing units. Finally, the load balancing strategy distributes the incoming data among the available threads.

In this work, we aim at identifying and solving the issues raised by the dynamic adaptation of an SPE resource allocation while facing critical fluctuations in input rate and value distribution. Most existing SPEs integrate efficient automatic scheduling strategies designed to implement different objectives. For example, the STORM family includes RSTORM [28], TSTORM [39], and Stela [40] which respectively aim at finding the scheduling plan that reduces the number of active processing units, thus minimizing network traffic between processing units and avoiding processing bottlenecks due to input overload. To attain this goal, each strategy affects the scheduling plan so that data are processed with short latency. For example, in TSTORM [39], authors highlight overheads generated by network communications. This observation is reused in [28] and extended to resource usage to define an optimal scheduling plan, *i.e.,* a scheduling plan involving minimal computation overheads. In this paper, we focus on parallelism degree and load balancing management so as to propose a solution that is compatible with each of the scheduling strategies. Our goal is to obtain a preventive solution which adapts the system to data stream evolutions before problems occur. Furthermore, we expect as general a solution as possible, which does not depend on users whether for obtaining information from their experience or for triggering system adaptations. To reach this goal, we propose to build over two already published solutions, AUTOSCALE [24] and OSG [29–31]. AUTOSCALE proposes a method to fix the parallelism degree of each operator of a workflow with an original data-driven approach, which considers the complete graph of operators and data streams in the workflow to avoid inconsistent local decisions that lead to rapid revisions and therefore significant system instability. An example of this is an operator starting a scale-in while it is apparent that its activity will augment soon due to an increase of the output stream(s) of upstream(s) operator(s). Unfortunately, AUTOSCALE presents some instability problems which had to be studied and fixed. OSG (ONLINE SHUFFLE GROUPING) deals with load balancing. Even if tuple processing times are not similar from one value to the other, OSG aims at reducing tuple completion times by carefully scheduling each incoming tuple.

The original contributions presented in this paper are three-fold:

1. An auto-parallelization strategy improving the approach presented in [24]. AUTOSCALE+, thanks to a better modeling of STORM effective resource usage, enables quicker deployment of adequate resources, thus improving system throughput and stability.
2. The integration of AUTOSCALE+ and OSG into DYNAMIC AND AUTOMATIC BALANCED SCALING FOR STORM (DABS-STORM), a holistic and automatized approach to parallelism and load balancing in stream processing systems, has been enabled due to their compatibility.
3. A thorough experimental evaluation of DABS-STORM highlighting its ability to process streams with critical fluctuations in input rate and value distribution for complex continuous queries. In addition, we compare DABS-STORM with well-known approaches from the literature.

In the remainder of this paper, Sect. 2 presents the execution context from logical and physical points of view. We describe how continuous queries are processed over distributed infrastructures and the processing model. Section 3 presents the related work, reviewing the background on dynamic and elastic stream processing and the main elasticity mechanisms at infrastructure and query levels for handling variance in input load. Approaches for parallelism management and load balancing are described, respectively, in Sects. 4 and 5. Our original approach, DABS-STORM, is detailed in Sect. 6 while Sect. 7 is devoted to its experimental evaluation.

2   System Model

2.1   Execution Environment

To make things more concrete while introducing some notations, let us consider three continuous queries Q_1, Q_2 and Q_3 represented by workflows $W1$, $W2$ and

$\mathcal{W}3$ with respective associated output streams S'$_1$, S'$_2$ and S'$_3$ (see Fig. 1). A workflow $\mathcal{W} = (\mathcal{O}, \mathcal{V})$ is a direct acyclic graph where $\mathcal{O}$ is the set of operators and $\mathcal{V}$ the set of streams. Presented workflows are quite simple: $\mathcal{W}1$ is linear, $\mathcal{W}2$ is a diamond, while $\mathcal{W}3$ is a star. Despite their simplicity, these workflows are interesting to study. Indeed, they are general patterns used to build much more complex workflows [28]. Each workflow processes a set of input stream(s) which, in our example, is included in $\{S_1, S_2, S_3\}$.

A stream is a potentially infinite sequence of *tuples*, *i.e.*, key/value pairs, arriving over time. An input stream may have fluctuations in input rate and value distribution as shown on the left of Fig. 1. It is worth noting the impact these fluctuations can have, not only on the processing time, but also on the *selectivity* of operators, *i.e.*, the ratio between the number of output and input tuples. This second point can be critical for operators such as joins [15,37] as well as having direct impact on downstream operators.

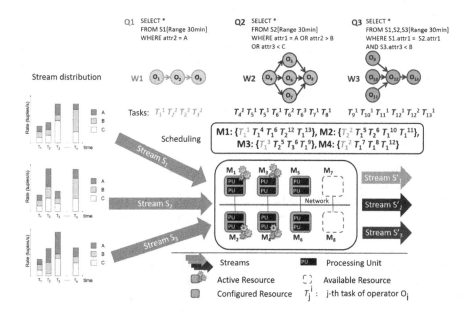

Fig. 1. Distributed stream processing.

Each operator $O_i \in \mathcal{O}$ is processed in parallel. The parallelism degree $d(O_i)$ of operator O_i corresponds to the number of tasks assigned to the operator. For instance, on Fig. 1, operator O_2, executed by tasks T$_2^1$ and T$_2^2$, has a parallelism degree of $d(O_2) = 2$.

A scheduling strategy assigns tasks to the processing units, in this case eight available machines (M$_1$ to M$_8$). For instance, on Fig. 1, the four tasks of the workflow $\mathcal{W}1$ are distributed on machines M$_1$ to M$_4$.

For a machine, three states are possible: *active*, *configured* and *available*. On Fig. 1, machines M$_1$ to M$_4$ are active, and run some assigned tasks. Machines

M_5 and M_6 are configured but inactive (no assigned tasks). Finally, M_7 and M_8 are available but not configured, which means the scheduler cannot assign tasks to them.

While DABS-STORM could be extended to handle heterogeneous machines, in this paper, for the simplicity's sake, we restrict the execution context to homogeneous machines (**H1**), *i.e.*, all machines have the same amount of CPU, RAM and bandwidth. Furthermore, we also assume that there is never resource starvation. In other words, there are always enough computational resources available to process input streams, no matter their input rates (**H2**). Thus, a loss of quality or performance cannot be ascribed to a lack of resources. To enforce **H2**, if the scenario does not ensure it, load shedding techniques [22,30] can be relied on, which drop some of the inputs in order to prevent buffer overflows or trashing.

Stream elements are assumed to be heterogeneous with respect to processing latency, depending on their values (**H3**). Consequently, DABS-STORM can handle homogeneous data streams as well as heterogeneous ones.

Finally, as we aim at proposing a generic solution supporting user-defined functions as well as common operators like filters and joins, we intend to deal with stateless and stateful operators. However, in this paper, we focus on stateless operators (**H4**). Indeed, it has been shown that solutions can also be useful for other kinds of operators [31], and most SPEs supporting stateful operators, like joins, provide a state management method while replicating these operators, such that we can rely on them for this part.

2.2   Processing Model

Each operator O_i has a logical input stream $\sigma_i = \langle e_1, \ldots, e_q, \ldots, e_m \rangle$. Since operator O_i may be executed in parallel by $k = d(O_i)$ tasks $T_i^1, \ldots, T_i^k$, then each task receives a physical input sub-stream $\sigma_i^1, \ldots, \sigma_i^k$. Notice that $\sigma_i = \bigcup_{x \in [k]} \sigma_i^x$. Tuples of σ_i are assigned to a sub-stream, and thus to a task, according to a predefined load balancing strategy. We denote by $f(e)$ the unknown frequency[1] of tuple e, *i.e.*, the number of occurrences of e in the stream of size m. Before being processed, a tuple e_q is buffered in a FIFO input queue consumed by a task. The processing latency $w_i^x(q)$ of tuple e_q on the task T_i^x depends on the time complexity of O_i, on the computational power available to task T_i^x, and potentially, on the values of e_q attributes. Without loss of generality, we assume that tuples in a stream σ are identified by a single integer drawn from a large universe $[n] = \{1, \ldots, n\}$. In other words, tuples can be modeled as single values. The processing latency is modeled as an unknown function[2] of the value of e_q. The probability distribution of e_q values may vary over time. In a stable system the average processing latency of operator O_i can be defined as

$$\overline{W}_i = \frac{1}{|\sigma_i|} \sum_{x \in [k]} \sum_{e_q \in \sigma_i^x} w_i^x(q) \tag{1}$$

---

[1] This definition of *frequency* is compliant with the data streaming literature [7,35].
[2] The experimental evaluation relaxes the model by taking into account processing latency variance.

Let $\ell(q)$ be the completion time or end-to-end latency of e_q, *i.e.*, how much time it took for tuple e_q from the instant it was inserted into the assigned task's buffer to when it was processed by the tasks. Then we can define the average completion time for operator O_i as

$$\overline{L}_i = \frac{1}{|\sigma_i|} \sum_{e_q \in \sigma_i} \ell(q) \qquad (2)$$

Table 1 summarizes the notation.

Table 1. Notations.

Workflow/topology	$\mathcal{W}$
Workflow input and output streams	S, S'
Operator	$O_i \in \mathcal{O}$
Parallelism degree	$d(O_i)$
Task of operator O_i	$T_i^x, x \in [k]$
Task T_i^x and operator O_i input streams	$\sigma_i^x \subseteq \sigma_i$
q^{th} tuple in the stream σ_i	$e_q \in \sigma_i$
Processing latency of e_q on tasks T_i^x	$w_i^x(q)$
Average processing latency of operator O_i	$\overline{W}_i$
Completion time of e_q	$\ell(q)$
Average completion time of operator O_i	$\overline{L}_i$
Tuple e frequency	$f(e)$
Tuple e empirical probability of occurrence	$p(e)$
Size of the stream	m
Number of distinct tuples in the stream	n
2-universal hash function	h

3   Related Works

This section presents and discusses the most relevant strategies in the literature. Adaptation mechanisms aiming at maintaining processing within some performance goals are said to be *elastic* [32], *i.e.*, they adapt to input stream variance. Considering the huge difference between elastic mechanisms working at physical level (i.e. adapting resource consumption at infrastructure level) and those working at logical level (elastic mechanisms adapting workflows to fit processing load requirements), in this paper we only focus on the latter.

3.1 Elastic Mechanisms at Logical Level

Workflows can be adapted to handle variations in input load. Logical level approaches can be classified as parallelism management, scheduling, and load-balancing.

Parallelism Management—To process stream elements timely, operator output throughput should be greater than input throughput (taking into account also the selectivity factor). Nevertheless, with a fixed number of threads, two critical cases can occur:

- If input throughput is greater than output throughput for a sizable time period, then the number of elements in the buffering queue increases. This scenario causes an unacceptable increase in end-to-end latency and may lead to *congestion* [22,33].
- If input throughput is smaller than output throughput for a sizable time period, then buffering queues are mostly empty and tasks are often idle. While in this scenario the system has low latencies, it also implies that resource usage is not maximized.

To handle these critical scenarios, SPEs should integrate a more refined parallelism management strategy. When facing an overload, SPEs should increase the parallelism degree (*scale-out*) of operators, thus decreasing the queuing time of incoming elements. Conversely, when input throughput is low, SPEs should decrease the parallelism degree (*scale-in*) to minimize resource waste.

Scheduling Strategy—Given the operator parallelism degree, SPEs must schedule the tasks to the available processing units (Fig. 1). We identify three classes of scheduling strategies:

- Strategies based on CPU load balancing between all processing units [1,27,41] assign threads on as many units as possible to divide processing load evenly. Using all available resources is an appropriate solution to limit processing bottlenecks due to CPU shortage. The problem is that it may imply massive network overheads [39] and underused units.
- Strategies based on network traffic reduction [3,39] tend to concentrate as many threads as possible on the same processing units to minimize network traffic. These approaches improve throughput of SPEs [39] and reduce the number of active machines compared to the previous class. However, when input rate increases significantly, active machines tend to be overloaded more quickly and imply major reconfiguration compared to strategies spreading load evenly among all available units.
- Resource-aware strategies [3,28] aim at avoiding processing unit overload and minimizing resource consumption. Through resource monitoring and processing requirements, this class of scheduling strategies allows threads to be grouped on processing units, thus minimizing resource waste. It offers efficient scheduling while having resource requirements for each thread to be assigned. The problem is that it requires accurate specifications about resource requirements and thus relies on user expertise. If user specifications are oversized or undersized, this leads to a waste or lack of resources, respectively.

Intra-operator Load Balancing—Operators can be classified as being either *stateful* (*e.g.,* standard deviation computation) or *stateless* (*e.g.,* filtering).

When the target operator is stateful, its state must be kept continuously synchronized among its instances, with potentially severe performance degradation at runtime; a well-known workaround to this problem consists in partitioning the operator state and letting each instance work on the subset of input stream containing the tuples affecting its state partition [22]. In this case, *key grouping* is the preferred choice as stream partitioning can be performed to correctly assign all the tuples containing specific data values and only those to the same operator instance, thus greatly simplifying the development of parallelizable stateful operators at the expense of performance.

In recent years there has been new interest in improving load balancing with key grouping [17,26,31]. It is worth noting that all works cited assume that all tuples of a stream have the same execution time.

Considering stateless operators, *i.e.,* data operators whose output is only a function of the current tuple in input, parallelization is straightforward. The grouping function is free to assign the next tuple in input stream to any available instance of the receiving operator (contrary to stateful operators, where tuple assignment is constrained). Such stream partitioning functions are often called *shuffle grouping* and represent a fundamental element of a large number of stream processing applications [22]. Notice that solutions for shuffle grouping techniques can be applied to stateful operators as well, provided that the operator implementation includes some mechanism to warranty state consistency (*e.g.,* a subsequent reduce phase). Given its generality, in this work we consider only shuffle grouping stream partitioning.

Typical implementations of shuffle groupings are based on round-robin scheduling [4,5]. However, the processing latency of many operators are intrinsically sensitive to values. For example, an operator applying a transformation on each character of a text has a processing latency depending on the length of the text. Thus, high fluctuations in such values most likely increase load imbalance considerably, which lead to performance problems.

3.2 Triggering Elastic Stream Processing

Solving operator congestion in a stream processing context is a complex problem. Out of the three major factors (parallelism management, scheduling, load balancing), to our knowledge, most works [3,18,33,39] address only one at any time.

However, a clear distinction can be made between *reactive* approaches [21, 33,40], which detect and remove potential problems from the current state of the system, and *proactive* approaches which predict potential problems and anticipate solutions [15,34]. Among reactive solutions, we distinguish between *on user-demand* [40] and *automatic* [21,33] solutions.

In [40], authors suggest a solution triggering *scale-in* and *scale-out* on user demand. This solution relies on the user adding enough resources when throughput declines. Consequently, this solution is mainly limited by the need for user expertise and presence.

Dynamic and automatic approaches [18,33] also aim at adapting parallelism degrees to avoid congestion of operators. They are based on global and local consumption thresholds (CPU, memory), which aim at separating a normal consumption from a critical one. In addition, in [21], authors suggest an algorithm integrating a knowledge base, built through a learning phase and updated at runtime. This knowledge base associates parallelism degrees with expected throughput for each operator. These solutions share the fact of using current resource consumption to detect potential congestion, thus making anticipation almost impossible. Furthermore, they pay no attention to data distribution within the input data streams.

Finally, some model-based solutions [15,34] anticipate congestion, thanks to a complete model of the execution support and operator features (processing latencies, pending queues, *etc.*). Here, the parallelism degrees are adapted to minimize overall latency. Unfortunately, these solutions require detailed characteristics of the system and do not support any evolution of the execution support. In [38], authors suggest the Chronostream system, which is able to scale operators transparently and to manage internal states for both stateless and stateful operators. Even if this approach has demonstrated its efficiency in terms of scalability, Chronostream relies on stream partitioning to balance the load between operator instances. Thus, if there is a significant difference between the average processing latency for distinct keys, Chronostream is unable to compensate that imbalance accurately.

Summarizing SPE elasticity at logical level, the elasticity of a SPE mainly depends on choices related to parallelism management, scheduling, and load-balancing. Other aspects like workflow optimization [2,6] and implementation selection [22] are user-provided and cannot be modified at runtime. In this context, we aim at suggesting a stream-based solution scaling treatments according to stream evolution in terms of input rate and value distribution.

4 Parallelism Management with AUTOSCALE+

In [24] we defined a proactive approach, named AUTOSCALE, to manage dynamically and automatically the parallelism degree of operators using indicators monitored on streams and operators. Our proposed algorithm decides which operators have to be reconfigured (scale-out or scale-in) and what their new parallelism degrees are. These decisions are based on estimations of data stream evolution and resource consumption, which are computed from monitored indicators. The main originality of AUTOSCALE is that it considers the workflow as a whole, and more precisely the dependencies between operators, when validating a reconfiguration decision. It is worth noting that the algorithm we proposed offers satisfying results in deciding when a reconfiguration is required, but that the new parallelism degree computed was not always relevant, generating too frequent reconfiguration, thus leading to system instability in some specific cases. We have investigated the reasons for such behaviors and identified two main causes. The first corresponds to variations in computation times from one item to another depending on their values. Clearly, variations in distribution of these values within the input streams also has an impact. The problem is that such variations of computation times jeopardize the default STORM load balancing method. To solve this problem, the only solution is to replace the load-balancing method with a new method that has to pay attention to such variations (see Sect. 5). The second cause is related to the AUTOSCALE method itself. It transpired that the resource consumption analysis was not precise enough. In AUTOSCALE+, it has to be improved to better fit the specificities of the STORM's architecture. In this section, we first recall the general principles of AUTOSCALE before describing new methods embedded in AUTOSCALE+. This new proposal improves AUTOSCALE, taking CPU usage and user constraints into account.

4.1 AUTOSCALE+ Metrics

After presenting the general principles, we focus on the monitoring problem. Finally, we detail the new metrics embedded in AUTOSCALE+.

Principle
Just like its predecessor AUTOSCALE [24], AUTOSCALE+ anticipates potential congestion of operators through stream and operator monitoring[3]. For each operator, input volumes in the near future are estimated according to time series analysis [10]. The two algorithms also share the analysis of many dynamic properties like processing latency, pending queues, and the selectivity of each operator. Based on these, processing rates, or *capacity* can be estimated. The combination of these estimations make it possible to recommend scale-in, scale-out or nothing for each operator. Depending on available and configured resources dedicated to the SPE, a reconfiguration of threads running on the cluster could be triggered.

In contrast to AUTOSCALE, AUTOSCALE+ considers precise resource usage in terms of CPU, RAM and bandwidth. This allows improvement of decisions, for example avoiding reconfigurations when parallelism degree is not the root cause of problems, and more quickly reaching the adequate parallelism degree.

Monitoring Management
Monitoring management is based on sliding windows observing simultaneously all threads assigned on the execution support.

Let $\mathcal{F}$ be a set of monitoring sliding windows $\mathcal{F}_i = \{(F_j^i)\}_{j \in \mathbb{N}^+}$. Each window $\mathcal{F}_i$ is associated with an operator $\mathcal{O}_i$, and is composed of iterations F_j^i. Each iteration F_j^i is defined by a duration Δ and gathers measurements collected during this interval. These measurements are collected according to a predefined set of timestamps $\mathcal{M}_i = \{m^{i,1}, m^{i,2}, \ldots, m^{i,n}\}_{n \in \mathbb{N}^+}$. For each operator $\mathcal{O}_i$, our approach collects some measurements taking into account the stream elements

---

[3] At each scale-in or scale-out, system monitoring is disabled while the system stabilizes. Indeed, the data acquired during this transition period do not provide any information about the nominal behavior of the new configuration of the system.

received and processed on the interval $[m^{i,p-1}, m^{i,p}[$ with $p \in [n]$ where $[n] = \{1, \ldots, n\}$. More information about monitoring management is available in [24].

Let $\mathcal{R}^i$ be the set, potentially infinite, of stream elements received by operator $\mathcal{O}_i$. We consider $\mathcal{R}^{i,j}$ as the subset of elements received by $\mathcal{O}_i$ during the iteration F_j^i, and $\mathcal{R}^{i,p}$ the subset of elements received between $[m^{i,p-1}, m^{i,p}[$.

Let us now consider a parent operator $\mathcal{O}_{par}$ and a child operator $\mathcal{O}_{ch}$ consuming stream elements produced by $\mathcal{O}_{par}$. For both operators, we observe inputs $\mathcal{R}^{i,j}$. These inputs are inserted in pending queues where elements are consumed by associated functions. We define as $Input^{i,j}$ the sum of inputs processed currently by the function and stream elements pending in the queue during the iteration F_j^i. At the same time, we monitor the processing latency of the function and its selectivity factor for filter-based operators as presented in [23].

Metrics on Operator Input and Output

From these monitored values, we compute some metrics to analyze the activity of each operator. The aim is to identify operators which could have critical input volumes according to their processing capacities in the near future.

To do this, incoming volumes during the next iteration of the monitoring window are estimated, see Fig. 2. This estimation, called $EstimR^{i,j}$, is computed using a regression function f_{j-1}^i computed based on the previous iteration as follows:

$$EstimR^{i,j} = \sum_{m_p^i \in \mathcal{M}_i} \lceil f_{j-1}^i(m_p^i) \rceil \tag{3}$$

where each m_k^i belongs to the next iteration of the window.

To estimate precisely f_{j-1}^i, Autoscale+ selects the best candidate, i.e. the one best fitting to the previous iteration, among three competitors: linear, logarithmic, and exponential regression models. Compared to Autoscale, the computation overhead is very small, while stream fluctuations are improved.

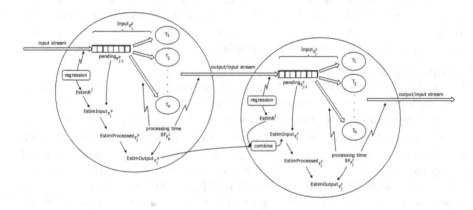

Fig. 2. Illustrating metrics considering two operators.

Operator history is not the only information that can be considered. Indeed, the near future of an operator is clearly influenced by antecedent operators. Furthermore, for an (antecedent) operator, a simple combination of the already computed $EstimInput^{i,j}$ with the average selectivity factor $\overline{\alpha}(F^i_{j-1})$ provides an estimation of its output $EstimOutput^{i,j}$:

$$EstimOutput^{i,j} = EstimInput_{F^i_j} \times \overline{\alpha}(F^i_{j-1}) \tag{4}$$

So, considering an operator who has ancestors, to estimate its inputs in a near future, we have two pieces of information. On the one hand, we have $EstimInput^{i,j}$ computed from its history, while, on the other hand, we have $EstimOutput^{i,j}$ the estimated outputs of its antecedent operators. A *combine* strategy is used to mix these elements. Last but not least, to approximate the total volume of stream elements each operator will have to process during the next iteration, attention must be paid to the pending queue. Stream elements pending in the operator queue, noted $pending^{i,j-1}$, just have to be added to the estimation. Finally, $EstimInput^{i,j}$ is defined as:

$$EstimInput^{i,j}$$

$$= combine\left(EstimR^{i,j}, \sum_{p\in par(\mathcal{O}_i)} EstimOutput^{p,j}\right) + pending^{i,j-1} \tag{5}$$

where $par(\mathcal{O}_i)$ returns all parent operators of $\mathcal{O}_i$.

Many different *combine* functions can be proposed or obtained by learning techniques. By default, AUTOSCALE+ simply returns the *max* of the two values. This corresponds to a cautious strategy with respect to scale-in operation. Indeed, scale-in is analyzed with respect to the highest estimation. On the contrary, the *combine* strategy can return the *min* estimation to avoid overconsumption of resources due to an ephemeral increase in input rates. The strategy used will depend on the user's priorities.

Operator Capacity Estimation
Intuitively, the capacity of an operator to treat items during a period Δ can be estimated considering the processing time of elements.

$$IdealCapacity^{i,j} = \frac{\Delta}{Lat^{i,j}} \tag{6}$$

where $Lat^{i,j}$ is the processing latency.

This approximation would be quite good if computational resources used by an operator were constant. Unfortunately, it is not so simple. For example, a task can take advantage of free CPU to make use of more CPU than reserved. To illustrate this point, let us consider an example of three operators $\mathcal{O}_A$, $\mathcal{O}_B$ and $\mathcal{O}_C$, executed, respectively, by threads T_A, T_B and T_C, running on a single CPU, C. As depicted on Fig. 3, some reservations have been made for each of them [28], let's say $ResaCPU_A$, $ResaCPU_B$ and $ResaCPU_C$. While the interest of this constraint is to avoid assignments leading to resource starvation, it should be kept in mind that a resource used by a task is not fully defined by the reservations made for it.

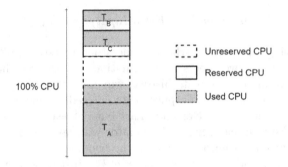

Fig. 3. Usable CPU for threads on one core

Considering Fig. 3, threads T_B and T_C use less CPU than they reserved. Clearly, there is no congestion since some CPU time is not used. Thread T_A takes advantage of the situation.

Here, the problem is to estimate usable CPU in a near future, and this has to be done for each operator/thread. We estimate the usable CPU by a thread T_X according to formula 7, where a *weighting factor* $\lambda \in [0;1[$ has been introduced to underestimate slightly available CPU and thus avoid fast overload.

$$UtilCPU(T_X, C) = \lambda \times max(UsedCPU(T_X, C), ResaCPU_X)$$
$$+ \frac{1}{n}(100 - \sum_{\forall T_Y \neq T_X} UsedCPU(T_Y, j)) \quad (7)$$

Considering an operator $\mathcal{O}_i$, to estimate the CPU it can use ($UtilCPU_i$), all its associated threads have to be considered ($T_i^1, T_i^2, \ldots, T_i^m$). The global CPU time $UtilCPU_i$ for $\mathcal{O}_i$ is estimated as follows:

$$UtilCPU_i = \min_{x=1\ldots x=m} (UtilCPU(T_i^x, CPU(T_i^x)) \quad (8)$$

Here, we assume that the input rate increase is spread evenly across all threads.

Capacity is then estimated:

$$Capacity^{i,j} = \frac{\Delta}{Lat^{i,j}} \times UtilCPU_i \quad (9)$$

4.2 Parallelism Degree Management in AUTOSCALE+

We now have enough information to detect imbalances between processing requirements and resource usage. Analysis of divergences will lead to one of the following three conclusions: need for scale-out, possibility of scale-in, or do nothing. If scale-out is often a need, scale-in is only a possibility. Indeed, these two operations are costly, and system stability is important to avoid wasting computing resources and time.

An "Ideal" Parellism Degree

Considering the estimations of the incoming workload and of the capacity, the ideal parallelism degree, leading the operator to efficient stream processing and denoted $idealK$, can be estimated according to:

$$idealK = \frac{EstimInput^{i,j}}{Capacity^{i,j}} \tag{10}$$

A Working Interval for Stability Issues

Stability is a major issue for an automatic process and it is important to find a good balance. As scale-outs are needed for the system to work correctly, the focus is on scale-in operations. Intuitively, even if it is possible, the scale-in operation will be retained until a large benefit is attained. To encode this intuition, we introduce a "working interval": this defines a zone where, even if the estimated $idealK$ is smaller than the actual parallelism degree, no reconfiguration will be carried out due to a lack of benefits. Furthermore, we suggest a controllable interval: its size should vary depending on the parallelism degree, and it should be controllable with respect to different considerations (user preferences, evolution -e.g. reducing- over time and so on). Trivially, to define an interval, two bounds have to be defined. The upper bound will be the current parallelism degree, while the lower bound min_k will be a function of the current parallelism degree k:

$$min_k = \beta \times k \tag{11}$$

where $\beta \in]0;1]$ is a controllable parameter. If β is close to 0, it means that AUTOSCALE+ performs scale-in only when input volumes are very small compared to operator capacities. If β is close to 1, AUTOSCALE+ performs scale-in as soon as possible.

It is worth noting that the greater k, the greater the associated working interval. We choose this property because the more tasks there are to merge, the more time it takes to merge pending queues distributed over the cluster and to re-route stream elements.

Decision of Parallelism Degree Modification

A scale-out should be performed for an operator $\mathcal{O}_i$ when $idealK$ exceeds the working interval as represented on Fig. 4.

On the other hand, if $idealK$ is smaller than min_k, a scale-in will be performed.

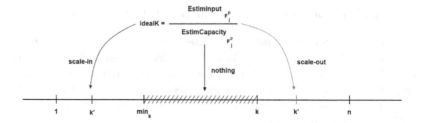

Fig. 4. Modification of parallelism degree

Otherwise, if *idealK* remains within the interval, even if a scale-in is possible, the operator is not modified and retains the current parallelism degree.

Computation of the Appropriate Parallelism Degree
Thus, each time a change is decided, whether it is a scale-in or a scale-out, AUTOSCALE+ computes a new appropriate parallelism degree k' according to estimated needs and computing resources possibilities:

$$argmin_{k'}\left(\frac{EstimInput^{i,j}}{ResCapacity^{i,j}} \leq k\right) \tag{12}$$

where particular attention must be paid to denominator. Indeed, $ResCapacity^{i,j}$ deals with resources, but compared to *idealK* there is a major difference: reservations $ResCPU_i$ are used instead of resource estimations. Indeed, such a change may lead to reallocations of threads and the only guarantee we have is the amount of resource required by the reservation. This leads to the formula:

$$ResCapacity^{i,j} = \frac{\Delta}{Lat^{i,j}} \times ResCPU_i \times \lambda \tag{13}$$

where $\alpha \in \ ]0;1]$ is a parameter allowing AUTOSCALE+ to consider a relative margin between effective CPU usage and CPU reservation. This means that AUTOSCALE+ takes into account the fact that some threads may need more than their reservation at runtime. Just like β, the parameter λ can be defined using several methods like empirical study, reinforcement learning or user expertise.

5   Load Balancing with OSG

While an adequate parallelism degree is important, alone it does not fully solve the problem. Indeed, changing the parallelism degree is not enough to address any variations in value distribution when facing a significant heterogeneity in terms of computational resource needs. Failure to pay attention to load balancing, as round-robin scheduling would do, may lead to imbalance problems (see Fig. 5).

Furthermore, imbalance problems could jeopardize any method trying to manage the parallelism degree, as AUTOSCALE+. A careful load balancing strategy, compatible with our philosophy and proposed solutions, is definitely needed.

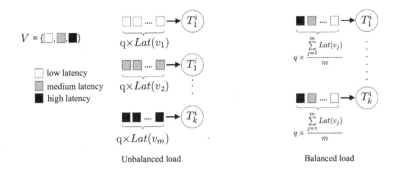

Fig. 5. Worst and optimal cases of load balancing

First, it has to be proactive. Prevention is better than cure since it avoids misleading the parallelism degree strategy. More generally, its behavior should not interfere with AUTOSCALE+ and both have to work together as smoothly as possible. Second, significant efforts have been made in AUTOSCALE+ to limit any dependence on the user. So as not to render these useless, the load balancing strategy should not require any user intervention.

This section is devoted to OSG, a load balancing strategy which has been specifically designed to deal with significant variations in computational resource needs depending on stream item values. We present its principles and main solutions. For more information, proofs and specific experimental evaluations, readers should refer to [29–31]. Indeed, this paper is focussed on the evaluation of DABS-STORM to study how AUTOSCALE+ and OSG interact and react to data stream fluctuations.

OSG is a shuffle grouping implementation based on a simple, yet effective idea: if we assume to know of the execution time $w_i^x(t)$ of each tuple t the parallel tasks of a given operator O_i, we can schedule the execution of incoming elements on such tasks with the aim of minimizing the average per tuple completion time of the tasks. However, the value of $w_i^x(t)$ is generally unknown. A common solution to this problem is to build a cost model for the execution time and then use it to pro-actively schedule the incoming load. However, building an accurate cost model usually requires a large amount of *a priori* knowledge on the system. Furthermore, once a model has been built, it can be hard to handle changes in the system or input stream characteristics at runtime.

To overcome all these issues, OSG takes decisions based on the estimation $\widehat{C}_i^x$ of the execution time assigned to task T_i^x of operator O_i, that is $C_i^x = \sum_{t \in \sigma_i^x} w_i^x(t)$. In order to do so, OSG computes an estimation $\widehat{w}_i^x(t)$ of the execution time $w_i^x(t)$ of each tuple t on task T_i^x of operator O_i. Then, OSG can also compute the sum of the estimated execution times of the tuples assigned to task T_i^x, i.e., $\widehat{C}_i^x = \sum_{t \in \sigma_i^x} \widehat{w}_i^x(t)$, which in turn is the estimation of C_i^x. A greedy scheduling algorithm (Sect. 5.1) is then fed with estimations for all the available operator tasks.

To implement this approach, each operator task builds a sketch (*i.e.*, a memory efficient data structure) that will track the execution time of the tuples it processes. When a change in the stream or task(s) characteristics affects the tuple execution times on some tasks, the concerned task(s) will forward an updated sketch to the scheduler that will then be able to (again) correctly estimate the tuples execution times. This solution does not require any *a priori* knowledge of the stream composition or the system, and is designed to continuously adapt to changes in the input distribution or the tasks load characteristics. Moreover, this solution is *proactive*, namely its goal is to avoid imbalance through scheduling, rather than detecting the unbalance and then attempting to correct it. A *reactive* solution can hardly be applied to this problem, as it would schedule input tuple on the basis of a previous, possibly stale, load state of the operator tasks. Furthermore, reactive scheduling typically imposes a periodic overhead even if the load distribution imposed by input tuples does not change over time.

For clarity's sake, we consider a topology with two operators: a non parallelized operator O_{sched} (*i.e.*, a *scheduler*), which schedules the tuples of a stream σ_{op}, and an operator O_{op}, whose k instances consume the stream σ_{op} (see Fig. 6). To encompass topologies where the operator generating the stream σ_{op} is itself parallelized, we can easily extend the model by taking into account parallel tasks of the operator O_{sched}. More precisely, there are s tasks/schedulers $T^1_{sched}, \dots, T^s_{sched}$, where task/scheduler T^x_{sched} schedules only a subset of σ_{op}, *i.e.*, its own output. In [31] we also show that in this setting OSG performances are better than the Round-Robin scheduling policy. In other words, OSG can be deployed when the operator O_{sched} is parallelized. Notice that our approach is hop-by-hop, *i.e.*, we consider a single shuffle grouped edge in the topology at a time. However, OSG can be applied to any shuffle grouped stage of the topology.

5.1 Count Min Sketch Algorithm

In [14], Cormode and Muthukrishnan introduced the Count Min sketch that provides, for each item e in the input stream, an (ε, δ)-additive-approximation $\hat{f}(e)$ of the frequency $f(e)$.

An algorithm is said to be an (ε, δ)-additive-approximation of the function ϕ on a stream σ if, for any prefix of size m of items of the input stream σ, the algorithm output $\hat{\phi}$ is such that $\mathbb{P}\{|\ \hat{\phi} - \phi\ | > \varepsilon C\} < \delta$, where ε, $\delta > 0$ are given as precision parameters and C is an arbitrary constant. The parameter ε represents the precision of the approximation estimation. For instance $\varepsilon = 0.1$ means that the additive error is less than 10% and $\delta = 0.01$ means that this approximation will not be satisfied with a probability less than 1%.

The Count Min sketch consists of a two dimensional matrix Φ of size $r \times c$, where $r = \lceil \log(1/\delta) \rceil$ and $c = \lceil 2.7/\varepsilon \rceil$. Each row is associated with a different 2-universal hash function $h_i : [n] \to [c]$.

A collection $\mathcal{H}$ of hash functions $h : [n] \to [c]$ is said to be 2-universal if for every two different items $x, y \in [n]$, for all $h \in \mathcal{H}$, $\mathbb{P}\{h(x) = h(y)\} \leq 1/c$, which is the probability of collision obtained if the hash function assigned truly random values in $[c]$. Carter and Wegman [11] provide an efficient method for building large families of hash functions approximating the 2-universality property.

When the `Count Min` algorithm reads item e from the input stream, it updates each row: $\forall i \in [r], \Phi[i, h_i(e)] \leftarrow \Phi[i, h_i(e)] + 1$. Thus, the cell value is the sum of the frequencies of all the items mapped to that cell. Upon request of f_e estimation, the algorithm returns the smallest cell value among the cells associated with t: $\hat{f}_e = \min_{i \in [r]} \{\Phi[i, h_i(e)]\}$.

Fed with a stream of m items, the space complexity of this algorithm is $\mathcal{O}(\log[(\log m + \log n)/\delta]/\varepsilon)$ bits, while update and query time complexities are $\mathcal{O}(\log(1/\delta))$. The `Count Min` algorithm guarantees that the following bound holds on the estimation accuracy for each item read from the input stream: $\mathbb{P}\{|\hat{f}(e) - f(e)| \geq \varepsilon(m - f_e)\} \leq \delta$, while $f(e) \leq \hat{f}(e)$ is always true.

This algorithm can be easily generalized to provide (ε, δ)-additive-approximation of point queries $\mathcal{Q}_e$ on a stream of updates, *i.e.*, a stream where each item e carries a positive integer update value v_e. When the `Count Min` algorithm reads the pair $\langle e, v_e \rangle$ from the input stream, the update routine changes as follows: $\forall i \in [r], \Omega[i, h_i(e)] \leftarrow \Omega[i, h_i(e)] + v_e$.

Greedy Online Scheduler. A classical problem in the load balancing literature is to schedule independent tasks on identical machines minimizing the makespan, *i.e.*, the *Multiprocessor Scheduling* problem. In this paper, we adapt this problem to our setting, *i.e.*, to schedule *online* independent tuples on non-uniform operator instances in order to minimize the average per tuple completion time $\overline{L}$. Online scheduling means that the scheduler does not know in advance the sequence of tasks it has to schedule. The Greedy Online Scheduler algorithm assigns the currently submitted tuples to the less loaded available operator instance. In [31] we show that this algorithm is a $(2 - \frac{1}{k})$-approximation of an optimal omniscient scheduling algorithm, namely an algorithm that knows in advance all the tuples it will receive. Notice that this is a variant of the join-shortest-queue (JSQ) policy [25], where we measure queue length as the time needed to execute all the buffered tuples, instead of the number of buffered tuples.

5.2 Online Shuffle Grouping Design

Each operator O_{op} task instance T_{op}^x maintains two `Count Min` sketch matrices (Fig. 6A): the first, denoted by Φ_{op}^x, tracks the tuple frequencies $f_{t,\text{op}}$; the second, denoted by Ω_{op}^x, tracks tuples cumulated execution times $\Omega_{\text{op}}^x = w_{\text{op}}^x(t) \times f_{\text{op}}^x(t)$. Both `Count Min` matrices have the same sizes and hash functions. The latter is the generalized version of the `Count Min` presented in Sect. 5.1, where the update value is the tuple execution time when processed by the instance (*i.e.*, $v_t = w_{\text{op}}^x(t)$). The operator instance will update both matrices after each tuple execution.

The operator tasks are modeled as a finite state machine (Fig. 7b) with two states: START and STABILIZING. The START state lasts until the task has

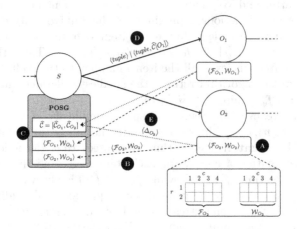

Fig. 6. OSG design where $r = 2$ ($\delta = 0.25$), $c = 4$ ($\varepsilon = 0.70$) and $k = 2$.

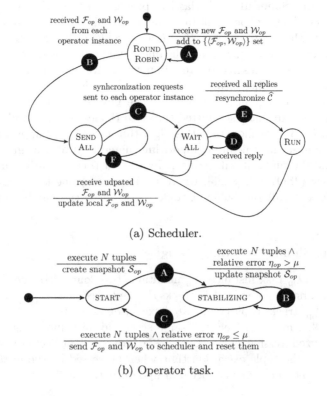

(a) Scheduler.

(b) Operator task.

Fig. 7. OSG finite state machines.

executed N tuples, where N is a user-defined window size parameter. The transition to the STABILIZING state (Fig. 7b Ⓐ) triggers the creation of a new snapshot Ψ_{op}^x. A snapshot is a matrix of size $r \times c$ where $\forall i \in [r], j \in [c]$: $\Psi_{op}^x[i,j] = \Omega_{op}^x[i,j]/\Phi_{op}^x[i,j]$. We say that the Φ_{op} and Ω_{op} matrices are stable when the relative error η_{op}^x between the previous snapshot and the current one is smaller than μ, that is if

$$\eta_{op}^x = \frac{\sum_{i=1}^r \sum_{j=1}^c \left| \Psi_{op}^x[i,j] - \frac{\Omega_{op}^x[i,j]}{\Phi_{op}^x[i,j]} \right|}{\sum_{i=1}^r \sum_{j=1}^c \Psi_{op}^x[i,j]} \leq \mu \tag{14}$$

is satisfied. Then, each time task T_{op}^x has executed N tuples, it checks whether Eq. 14 is satisfied. (i) If not, then Ψ_{op}^x is updated (Fig. 7b Ⓑ). (ii) Otherwise the task sends the Φ_{op}^x and Ω_{op}^x matrices to the scheduler (Fig. 6 Ⓑ), resets them and moves back to the START state (Fig. 7b Ⓒ).

There is a delay between any change in the stream or operator task characteristics and when the time the scheduler receives the updated Φ_{op}^x and Ω_{op}^x matrices from the affected operator tasks(s). This introduces a skew in the cumulated execution times estimated by the scheduler. To compensate for this skew, we introduce a synchronization mechanism that springs whenever the scheduler receives a new pair of matrices from any task. Notice also that there is an initial transient phase in which the scheduler has not yet received any information from operator instances. This means that, in this first phase, it has no information on the tuple execution times and is forced to use the Round-Robin policy. This mechanism is thus also needed to initialize the estimated cumulated execution times when the Round-Robin phase ends.

The scheduler maintains the estimated cumulated execution time for each task, in a vector $\widehat{C}$ of size k, and the set of pairs of matrices: $\{\langle \Phi_{op}^x, \Omega_{op}^x \rangle\}$, initially empty.

The scheduler is modeled as a finite state machine with four states: ROUND-ROBIN, SEND ALL, WAIT ALL, and RUN.

The ROUND-ROBIN state is the initial state in which scheduling is performed with the Round-Robin policy. In this state, the scheduler collects the Φ_{op}^x and Ω_{op}^x matrices sent by the operator tasks (Fig. 7a Ⓐ). After receiving the two matrices from each instance (Fig. 7a Ⓑ), the scheduler is able to estimate the execution time for each submitted tuple and moves to the SEND ALL state. When in the SEND ALL state, the scheduler sends the synchronization requests towards to the k tasks. To reduce overhead, requests are piggy backed (Fig. 6 Ⓓ) with outgoing tuples applying the Round-Robin policy for the next k tuples: the i-th tuple is assigned to operator instance $i \mod k$. On the other hand, the estimated cumulated execution time vector $\widehat{C}$ is updated with the tuple estimated execution time. When all the requests have been sent (Fig. 7a Ⓒ), the scheduler moves to the WAIT ALL state. This state collects the synchronization replies from the operator tasks (Fig. 7a Ⓓ). Operator task T_{op}^x reply (Fig. 6 Ⓔ) contains the difference Δ_{op}^x between the instance cumulated execution time C_{op}^x and the scheduler estimation $\widehat{C}[op]$.

In the WAIT ALL state, scheduling is performed as in the RUN state. When all the replies for the current epoch have been collected, synchronization is performed and the scheduler moves to the RUN state (Fig. 7a Ⓔ). In the RUN state, the scheduler assigns the input tuple applying the Greedy Online Scheduler algorithm, *i.e.,* assigns the tuple to the task with the least estimated cumulated execution time. Then it increments the target instance estimated cumulated execution time with the estimated tuple execution time. Finally, in any state except ROUND-ROBIN, receiving an updated pair of matrices Φ_{op}^x and Ω_{op}^x moves the scheduler to the SEND ALL state (Figure 7a Ⓕ).

Readers can refer to [29] for the complete theoretical analysis of OSG, in terms of correctness, accuracy and complexities.

6   DABS-Storm

We now have two methods, AUTOSCALE+, which adapts the parallelism degree of each operator, and OSG, which carefully balance streams' items between tasks of an operator. Integration seems quite easy. Nevertheless, mixing methods can always raise compatibility issues.

An auto-parallelization approach like AUTOSCALE+ assumes that congestion is due to an input overload of all tasks associated with an operator. In this case, adding more tasks is indeed the recommended solution. The better the load balance, the greater the scale-out effects will be. Furthermore, considering the definition of *UtilCPU* (see Eq. 8), the assumption of a good load balancing is present. Consequently, with a careful proactive load balancing preventing load imbalance, OSG is expected to improve both the decision process and the effects of AUTOSCALE+. OSG is not a random choice. Indeed, the proactive aspect is here of major importance. Although one might think that a reactive solution could work as well, sometimes (too often) a non-prevented load balancing problem could lead to unnecessary scale-outs of an unpredictable magnitude.

On the other hand, to guarantee good performance, OSG needs a non-congested environment. This means that scale-outs have to be performed before any congestion occurs. It also means that scale-in has to be performed cautiously, not too soon, to avoid any risk of congestion due to a reversal of the trend within data streams. AUTOSCALE is also a proactive method, and is designed to perform a scale-out before congestion occurs. For scale-in, things are less evident. First, as explained in Sect. 4.1, page 11, users can choose between two strategies (related to the choice of the *combine* function): a cautious one, or a resource-oriented one aimed at avoiding over-consumption of resources. Second, the size of the *working interval*, and more particularly the parameter β (see Sect. 4.2, page 14), is of major importance here. Indeed, it controls the margin of security thickness before performing a scale-in. To expect AUTOSCALE and OSG to work well together, the *combine* function must be a cautious one (i.e. a *max* function, which is the default choice), and the parameter β should be chosen close to zero to perform scale-in only when input volumes are very small compared to operator capacities.

To summarize, to take advantage of the benefits of both methods, it is not enough to use both of them. We must also ensure they are compatible. This seems to be the case here, provided that two AUTOSCALE+ parameters are correctly selected. However, an experimental study is essential to confirm this hypothesis.

7   Experiments

The implementations of AUTOSCALE+ and OSG which we use in experimental evaluations, have been developed to be integrated with Apache STORM [5]. The principles presented in AUTOSCALE+ and OSG could be integrated with many other DSMS, like for example Apache Spark Streaming [41], Flink [4], etc. However, for evident reasons of time and resources, at the beginning of the project we had to make a choice. We settled on STORM mainly for three reasons. First, in the STORM paradigm, stream elements must not be represented as key/value pairs, necessary for MapReduce-based approaches [16]. Second, it offers great flexibility for operator definition. Third, STORM serves as a guarantee that every item will be tracked and processed until an operator discards it (*e.g.* a filter or a final operator). Finally, it allows manual reconfiguration of parallelism degrees at runtime.

Thus, this section starts with some general reminders about this DSMS. Then we detail the experimental setups. Finally, we present and comment on the obtained results.

7.1   Overview of Apache Storm

Apache STORM [5] is an open source DSMS allowing users to express their continuous queries through a declarative language or to directly build their topologies using a high-level language (Java, Python, Clojure, etc.).

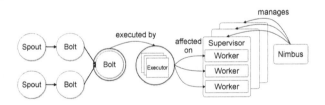

Fig. 8. Storm architecture

Whatever the language used, an operator, named a *component* in STORM's terminology, belongs to one of the two categories: *spouts* or *bolts*. A *spout* is a connector to a stream source, and thus can be used as an entry of a topology. It distributes stream elements to components to which it is connected and can process filtering operations if required. A *bolt* consumes items from any component and computes a result for each element received (*stateless* bolt) or for a set of elements (*stateful* bolt). Each *component* is executed in parallel by *executors*. Each *executor* is assigned to a processing unit by the scheduler (see Fig. 8).

7.2 Experimental Protocol

Experimental Setup. We experiment with the version 1.0.2 of Apache STORM. Our test cluster is composed of 10 VMs each with a dual-core CPU Intel(R) Xeon(R) E5-2620 running at 2.00 GHz, 4 GB of RAM and 40 GB of hard disk space. A master VM, called *Nimbus*, is responsible for coordinating the 9 others dedicated to task execution. Each of these VMs, called a *Supervisor*, manage 4 processing units, called *workers*. Our module, managing the operator parallelism degree, implements the *IScheduler* interface of the STORM API. The module, managing distribution of stream elements between executors, implements the *CustomStreamGrouping* interface of the STORM API. We also deploy a MySQL database on Nimbus to store collected measurements. We summarize the main experimental parameters in Table 2:

Table 2. Main parameters

Component	Description	Symbol	Value
STORM	Monitoring frequency		10 s
	Processing timeout		30 s
AUTOSCALE+	Weighting factor	λ	0.3
	Scale-in control	β	0.8
	Combine strategy		max
OSG	Precision	ϵ	0.05
	Maximal probability of error	δ	0.05

Test Topologies. To validate our approach, we demonstrate its effectiveness on three topologies.

(a) Simple insensitive topology.   (b) Simple sensitive topology.

Fig. 9. Simple topologies.

The simple insensitive topology (see Fig. 9a) composed of a spout (Source) emitting stream elements without filtering them. These stream elements are processed by a bolt (InsensitiveBolt) applying a function with a time complexity independent from the value read in input. Finally, a bolt (FinalizeBolt) ends the computation of each stream element by sending a termination signal to the STORM monitor.

The simple sensitive topology (see Fig. 9b) has the same structure as the simple insensitive topology. However, the function applied by the intermediate bolt (SensitiveBolt) has a time complexity that depends directly on the value read in the input.

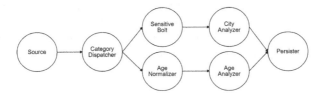

Fig. 10. Complex sensitive topology

The complex sensitive topology is inspired from real benchmark applications for Storm[4]. It is composed of several operators with various selectivity factors and average processing latencies. The spout (OpinionSource) emits stream elements concerning opinions submitted by users about a topic. Each opinion is described by information on the user, like her age and code representing her location, the topic and the user opinion. Stream elements are sent to a bolt (CategoryDispatcher) filtering unnecessary attributes and depending on the branch downstream. Moreover, it filters stream elements concerning a predefined list of irrelevant topics. A branch starts with a bolt (SensitiveBolt) retrieving information on user location from the code. This bolt has exactly the same properties as the sensitive bolt of the simple sensitive topology. Indeed, the time required to retrieve information on the city varies according to the code. It allows us to compare the impact of workflow structure and complexity on bolt behavior and dynamic adaptation of its parallelism degree. Then, a bolt (CityAnalyzer) extracts relevant subgroups according to opinion and location. The other branch, starting from the bolt CategoryDispatcher, performs similar treatments in order to define subgroups based on user opinion and age. Finally, the Persister takes as its input descriptions of subgroups and persists them in a storage file system.

It is important to bear in mind a major difference between this experimental setup and previous one and the consequences. The critical operator is not directly connected to the source. Stream elements are filtered and transformed by upstream operators. Consequently, its input rate may significantly differ from the source one, and strategies considering the workflow at a global scope will have a way to differentiate from those working only on local considerations. As a consequence, auto-parallelization strategies considering exclusively local observations cannot take advantage of fluctuations of input rate happening upstream. At the opposite, auto-parallelization strategies adapting parallelism degree of operators according to the global state of the workflow will have a natural advantage. So, this experimental setup highlights the fundamental difference between AUTOSCALE+ and other parallelization strategies considered in these experiments.

---

[4] https://github.com/yahoo/streaming-benchmarks.

Test Data Streams. As illustrated in Fig. 11, we build two synthetic streams with the following common features: (1) at least one critical increase in input rate leading the system to congestion with a minimal (one executor per operator) and static configuration (2) decrease in input rate to evaluate system elasticity. For each stream, we can set the distribution law, which may be uniform over all possible values or biased according to a zipf law with a predefined skew. These streams allow us to determine which impact of DABS-STORM when facing critical fluctuations in both input rate and value distribution. The first corresponds to a large increase with a plateau before a decrease. The second is much more sudden, with far severer variations, highlighting system elasticity.

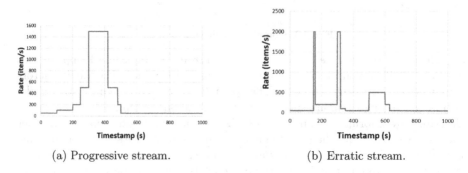

(a) Progressive stream.   (b) Erratic stream.

Fig. 11. Input streams.

Baseline Methods. First, we have to deplore a lack of open source implementation of auto-parallelization strategies. We compare DABS-STORM to two methods.

The first is simply the native static behavior of Apache STORM an incremental strategy (noted *incremental* hereafter) considering only thresholds on CPU usage.

The second [18], is a reinforcement learning-based strategy mapping input rates to appropriate parallelism degrees at runtime (noted *Rlearning* hereafter). For the experiments, the methods take advantage of a knowledge base base acquired through a training phase carried out using the test data streams. Then the knowledge base covers all the fluctuations encountered (which is not always the case in practice). More generally, this can be considered to be a favorable conditions.

7.3   Experiments and Results

Not all the experiments conducted are presented here. More information can be found on our companion website[5]. In addition, this website includes a comparison between AUTOSCALE and AUTOSCALE+. It gives an overview of the gap between

---

[5] https://dabs.liris.cnrs.fr.

the performance of the auto-parallelization strategy presented in [24] and results presented below. In the remainder of this section, the results described correspond to average values over 5 iterations for each configuration, thus lessening the impact of punctual anomalies during tests.

Variations in the Input Stream Rate over an Insensitive Topology. In this experiment, we confront the simple insensitive topology Fig. 9a with a stream with large, but not too erratic, variations in input rate, see Fig. 11a.

In this configuration, the volume of stream elements to process is the only impact factor and OSG is of little use. For the sake of equity, we choose to conduct an experiment where all compared solutions sharing the default grouping solution of STORM were denoted shuffle grouping. This means that here we only test the AUTOSCALE+ component of DABS-STORM. Note that other experimental evaluations show that, in this case, OSG behaves quite similarly to STORM's shuffle grouping.

As expected, the *reinforcement learning* strategy increases the parallelism degree of the observed operator *InsensitiveBolt* (Fig. 12a), before decreasing it just following the input rate. Nevertheless, these modifications have a major

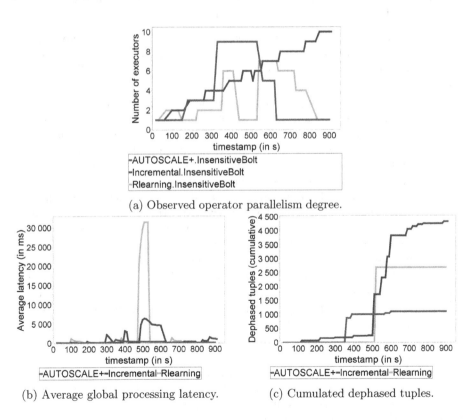

(a) Observed operator parallelism degree.

(b) Average global processing latency.   (c) Cumulated dephased tuples.

Fig. 12. Simple insensitive topology with progressive stream.

impact on average processing latency (Fig. 12b) and result quality (Fig. 12c). Indeed, the system has been reconfigured (scale-in) with respect to the input rate and without considering the pending queues (which were far from empty).

In comparison, the *incremental* strategy continuously increases the parallelism degree of the operator as long as the workload exceeds the processing rate (see Fig. 12). Even if the parallelism degree is increased, it cannot reach a suitable value as quickly as it needs to. This results in a large increase in average processing latency, causing a 29% loss of stream elements (dephased tuples) over the complete execution (Fig. 12c)). Moreover, in terms of resource usage, due to frequent system reconfigurations, the *incremental strategy* requires 64% more active processing units than the *reinforcement learning* strategy and 18% more than AUTOSCALE.

With AUTOSCALE+, Storm is able to anticipate suitable parallelism degrees over the complete execution. Even if AUTOSCALE+ tends to overestimate the required parallelism degree due to regression, processing latency is much better (Fig. 12b). It should also be noted that average processing latency remains remarkably stable, reducing losses to 7%.

We can conclude that AUTOSCALE+, thus DABS-STORM, outperform the baseline methods when confronted with data streams with large input rate variations, even if workflow is unresponsive to data values.

Erratic Variations in the Input Stream Rate over an Insensitive Topology. Compared to the previous experiment, the variations in the input stream rate will be more erratic, confronting the same insensitive topology Fig. 9a with the second data stream, see Fig. 11b.

As illustrated in Fig. 13a, the *reinforcement learning* strategy increases and decreases the parallelism degree of the observed operator *InsensitiveBolt* according to the two main peaks. However, the magnitude of the scale-out is not very high. Indeed, brief increases in input rate do not increase significantly the average input rate in recent history. Nevertheless, the, the sudden accumulation of a huge number of stream elements on pending queues increases the average processing latency. Luckily, the impact on result quality remains negligible with only 13% of stream elements lost over the complete execution as shown on Fig. 13c.

As the *incremental* strategy over-provisions the operator, available resources can hopefully handle brief increases in input rate. Consequently, the average processing latency (see Fig. 13b) increases significantly only when the input rate remains high over a long period of time such as for the last increase in the erratic stream. While stream element losses (Fig. 13c) are reduced to 19% over the complete execution, the usage of processing units remains higher than for the reinforcement learning strategy.

Considering Fig. 13a, AUTOSCALE+ reacts faster to sudden input rates increasing the parallelism degree. However, the increase is too high with respect to the ephemeral nature of the phenomenon. In other words, AUTOSCALE+ overestimates processing requirements. This overestimation induces excessive reconfiguration overheads, affecting punctually the average processing latency (Fig. 13b). Although results are delivered, 18% of the entire stream cannot be

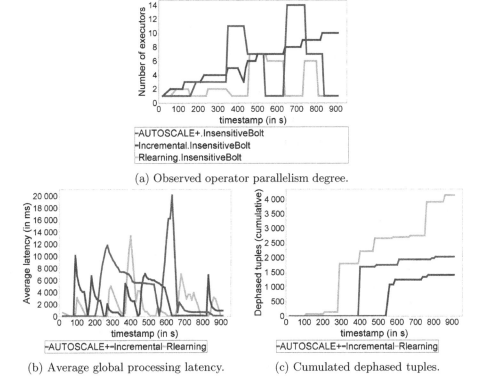

(a) Observed operator parallelism degree.

(b) Average global processing latency.   (c) Cumulated dephased tuples.

Fig. 13. Simple insensitive topology with erratic stream

processed under the maximal threshold. As critical increase and decrease in input rate are sudden and brief, they cannot be anticipated and affect processing latency before AUTOSCALE+ reconfigures the system.

This experimental setup points out the limit of the predictive approach when input streams vary suddenly in volume. Indeed, while a progressive evolution of the input rate can be easily anticipated, sudden peaks in input rate induce inappropriate behavior of the system behavior. Several solutions can be considered such as reinforcement learning that can help reduce this effect.

Variations in Input Stream Rate and Data Distribution over a Sensitive Topology. We now include variations in data distribution in the picture. The workflow and the data stream are of the same form as in experimentation Sect. 7.3 but with two major differences: first, in the input stream (Fig. 11a) the value distribution is biased, following a zipf distribution with a skew of 1.5[6]; and second, the workflow (Fig. 9b)) is sensitive to data values.

---

[6] This choice is motivated by previous results on OSG detailed in [31].

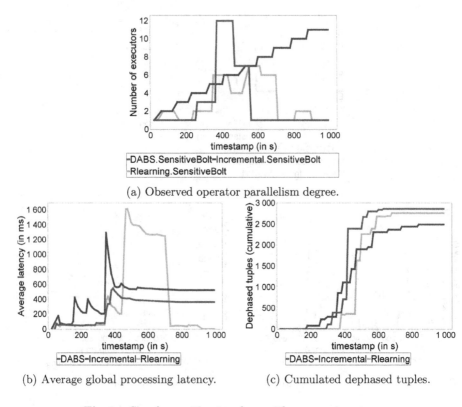

(a) Observed operator parallelism degree.

(b) Average global processing latency.

(c) Cumulated dephased tuples.

Fig. 14. Simple sensitive topology with progressive stream

For the sake of equity, in this experiment, each parallelism degree strategy is combined with OSG to benefit from a better load balancing.

Here, AUTOSCALE+ anticipates processing requirements (Fig. 14a) and is able to maintain a smaller processing latency (Fig. 14b), while the stream is at its maximal rate.

The *reinforcement learning* strategy is able to reduce processing latency significantly (Fig. 14b) when the input rate decreases.

With a parallelism degree evolution (Fig. 14a) very closely approaching that observed in experiment Sect. 7.3, the throughput of the *incremental* strategy is clearly less good. Having looked into this matter, this phenomenon is due to some kind of incompatibility between the incremental strategy and OSG. Indeed, step-by-step strategy trivially imposes frequent modifications of parallelism degree. At each step, OSG has to reevaluate its routing policy to keep a balance between executors.

In terms of tuple loss, all solutions deliver a similar performance even if the reinforcement learning strategy is able to keep losses under the incremental strategy and AUTOSCALE+. This is due to an overestimation of parallelism performed by AUTOSCALE+, affecting overall quality.

Concerning throughput, all solutions deliver a similar performance even if AUTOSCALE+ remains the auto-parallelization strategy keeping the smallest time shift between fluctuation in input rate and throughput.

Erratic Variations in Input Stream Rate and Data Distribution over a Sensitive Topology. Considering DABS-STORM, the average processing latency (Fig. 15b) remains low except for two punctual increases during the first two peaks in input rate and before the last increase in input rate which lasts longer. The loss of stream elements (Fig. 15c) is limited to 4.8%.

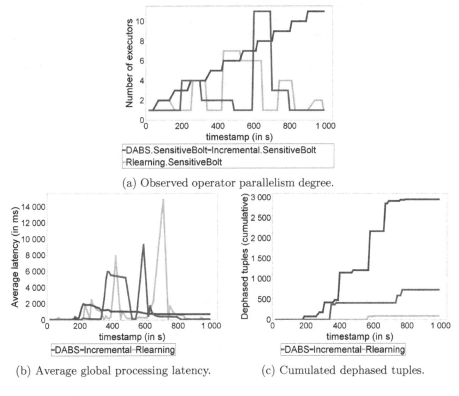

(a) Observed operator parallelism degree.

(b) Average global processing latency.   (c) Cumulated dephased tuples.

Fig. 15. Simple sensitive topology with erratic stream

The *reinforcement learning* strategy provides only the suitable number of executors (Fig. 15a) to avoid congestion.

While the *incremental* strategy maintains a low processing latency (Fig. 15b) and delivers a throughput close to the input rate, it uses considerably more resources to complete the treatment of the entire stream (Fig. 15a). Moreover, tuple loss (Fig. 15c) has the worst score of all three approaches.

It is also interesting to notice that DABS-STORM is quite reactive, maintaining a smaller time shift between fluctuations in input rate and throughput than the reinforcement learning solution. So, even if stream elements arrive at high rates, the proactive reconfiguration performed by DABS-STORM does not delay their treatment.

We observe with this configuration that DABS-STORM offers the best compromise between performance with moderate increases, in average, processing latency and acceptable losses.

Real Sensitive Topology Confronted with a Progressive Stream with Data Distribution Fluctuations. The topology used here, see Fig. 10, is representative of real-world continuous queries. It includes common operators such as filters on values and attributes, joins with static databases and also user-defined functions from expert domains such as data mining.

In such realistic conditions, DABS-STORM can take advantage of its global workflow approach. While a greater consumption of resources compared with other methods, see Fig. 16a, can be observed, it all other solutions in terms of processing latency, see Fig. 16b, and it also minimizes the loss of tuples, see Fig. 16c, while the parallelism degree clearly adapts to the input rate, see Fig. 16a.

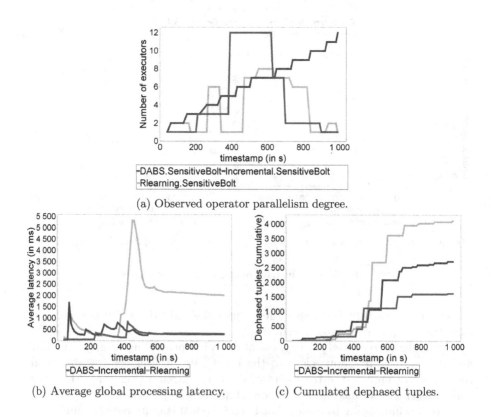

(a) Observed operator parallelism degree.

(b) Average global processing latency.   (c) Cumulated dephased tuples.

Fig. 16. Complex sensitive topology with progressive stream.

Real Sensitive Topology Confronted with an Erratic Stream. The *reinforcement learning* strategy keeps reacting to local average input rate to adjust the parallelism degree. It results in an inconsistent scale-in at workflow scope, which is contradicted afterwards with a major impact on processing latency and result quality as illustrated on Fig. 17b and c.

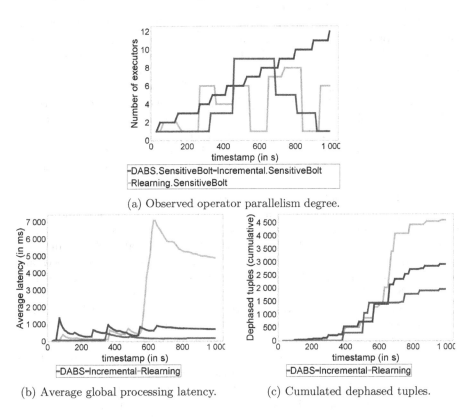

(a) Observed operator parallelism degree.

(b) Average global processing latency.   (c) Cumulated dephased tuples.

Fig. 17. Complex sensitive topology with erratic stream

Even when confronted with erratic stream variations again, considering both processing latency (Fig. 17b) and loss of tuples (Fig. 17c), we can conclude that DABS-STORM copes better than other solutions. This confirms the interest of not limiting the analysis to local considerations for each operator, but rather of having a data-driven approach to analyze the behavior on the entire workflow, as well as the complementarity and the compatibility of our two proposals, AUTOSCALE+ and OSG, making up DABS-STORM.

8 Conclusion

Proliferation and diversification of stream sources lead to new techniques in order to process large amounts of data with high velocity and quality. These techniques have to solve simultaneously three problems relating to management of the operators composing the workflow: parallelization, scheduling, and load balancing. In this paper, which focuses mainly on state-less operators, we have presented AUTOSCALE+, a proactive and coherent auto-parallelization strategy improving on AUTOSCALE [24]. It has been integrated into STORM with OSG, a cautious aware load balancing strategy, introducing a new member to the STORM family named DABS-STORM. Indeed, AUTOSCALE+ and OSG can be made perfectly compatible, complementing each other very well as two sides of the same coin. On the one side, OSG does not work well if the parallelism degree is underestimated, while AUTOSCALE anticipates to avoid such situation. On the other side, OSG improves load balancing, thus having a positive effect on the accuracy of estimations conducted by AUTOSCALE+ and reducing unnecessary reconfigurations. Such an agreement between two strategies is not systematic. For example, the experiments conducted highlight some incompatibility problems between OSG and the progressive strategy, which changes the parallelism degree one step at a time. Furthermore, DABS-STORM can be used in combination with different existing scheduling strategies.

Experimental evaluations have shown that DABS-STORM improves the system stability and performances. For example, even when facing brief and unpredictable fluctuations in input data streams, AUTOSCALE+ keeps loss of tuples under 18%. On complex real workflows, thanks to a global workflow analysis, DABS-STORM does even better, cutting losses to 10%. As long as the necessary resources are available, faced with large or very large fluctuations in input data streams, whether in terms of volume or data distribution (as can be observed, for example, in microblogging analysis), DABS-STORM is able to adapt. This automatic adaptation has many advantages. First, human supervision is no longer required to trigger and manage system reconfigurations. Note that a scarcity of resource remains a problem. Indeed, in the event of lack of resources blocking scale-out, DABS-STORM does not change its strategies as one would expect. One of our future research goals is thus to study the problem of resources starvation and to search for a solution that maximizes system throughput. Second, thanks to careful load balancing, dynamic parallelism degree adaptation with a global workflow analysis, and a data-driven approach, DABS-STORM manages computing resources better, reaching an interesting equilibrium between system stability and limited resource consumption.

References

1. Abadi, D.J., et al.: The design of the borealis stream processing engine. In: CIDR 2005, Second Biennial Conference on Innovative Data Systems Research, Online Proceedings, Asilomar, CA, USA, 4–7 January 2005, pp. 277–289. www.cidrdb.org (2005). http://cidrdb.org/cidr2005/papers/P23.pdf
2. Abadi, D.J., et al.: Aurora: a new model and architecture for data stream management. VLDB J. **12**(2), 120–139 (2003). https://doi.org/10.1007/s00778-003-0095-z

3. Aniello, L., Baldoni, R., Querzoni, L.: Adaptive online scheduling in storm. In: Chakravarthy, S., Urban, S.D., Pietzuch, P.R., Rundensteiner, E.A. (eds.) The 7th ACM International Conference on Distributed Event-Based Systems, DEBS 2013, Arlington, TX, USA, 29 June–03 July 2013, pp. 207–218. ACM (2013). http://doi.acm.org/10.1145/2488222.2488267
4. Apache Flink. https://flink.apache.org/
5. Apache Storm. https://storm.apache.org/
6. Arasu, A., et al.: STREAM: the Stanford data stream management system. In: Garofalakis, M.N., Gehrke, J., Rastogi, R. (eds.) Data Stream Management. DSA, pp. 317–336. Springer, Heidelberg (2016). https://doi.org/10.1007/978-3-540-28608-0_16
7. Arasu, A., Babu, S., Widom, J.: The CQL continuous query language: semantic foundations and query execution. VLDB J. 15(2), 121–142 (2006). https://doi.org/10.1007/s00778-004-0147-z
8. Balazinska, M., Balakrishnan, H., Stonebraker, M.: Load management and high availability in the medusa distributed stream processing system. In: Proceedings of the 2004 ACM SIGMOD International Conference on Management of Data, pp. 929–930. ACM (2004)
9. Biem, A., et al.: IBM infosphere streams for scalable, real-time, intelligent transportation services. In: Elmagarmid, A.K., Agrawal, D. (eds.) Proceedings of the ACM SIGMOD International Conference on Management of Data, SIGMOD 2010, Indianapolis, Indiana, USA, 6–10 June 2010, pp. 1093–1104. ACM (2010). http://doi.acm.org/10.1145/1807167.1807291
10. Box, G.: Box and Jenkins. In: Time Series Analysis, Forecasting and Control, pp. 161–215. Palgrave Macmillan, London (2013). https://doi.org/10.1057/9781137291264_6
11. Carter, L., Wegman, M.N.: Universal classes of hash functions. J. Comput. Syst. Sci. 18(2), 143–154 (1979). https://doi.org/10.1016/0022-0000(79)90044-8
12. Chandrasekaran, S., et al.: TelegraphCQ: continuous dataflow processing. In: Halevy, A.Y., Ives, Z.G., Doan, A. (eds.) Proceedings of the 2003 ACM SIGMOD International Conference on Management of Data, San Diego, California, USA, 9–12 June 2003, p. 668. ACM (2003). http://doi.acm.org/10.1145/872757.872857
13. Cherniack, M., et al.: Scalable distributed stream processing. In: CIDR 2003, First Biennial Conference on Innovative Data Systems Research, Online Proceedings, Asilomar, CA, USA, 5–8 January 2003. www.cidrdb.org (2003). http://www-db.cs.wisc.edu/cidr/cidr2003/program/p23.pdf
14. Cormode, G., Muthukrishnan, S.: An improved data stream summary: the count-min sketch and its applications. J. Algorithms 55(1), 58–75 (2005). https://doi.org/10.1016/j.jalgor.2003.12.001
15. Das, R., Tesauro, G., Walsh, W.E.: Model-based and model-free approaches to autonomic resource allocation. Technical report, RC23802, IBM Research Report, November 2005. http://domino.watson.ibm.com/library/cyberdig.nsf/1e4115aea78b6e7c85256b360066f0d4/f5e3b7f574b24bad852570c1005e35a9!OpenDocument&Highlight=0,tesauro
16. Dean, J., Ghemawat, S.: MapReduce: simplified data processing on large clusters. In: Brewer, E.A., Chen, P. (eds.) 6th Symposium on Operating System Design and Implementation (OSDI 2004), San Francisco, California, USA, 6–8 December 2004, pp. 137–150. USENIX Association (2004). http://www.usenix.org/events/osdi04/tech/dean.html
17. Gedik, B.: Partitioning functions for stateful data parallelism in stream processing. VLDB J. 23(4), 517–539 (2014). https://doi.org/10.1007/s00778-013-0335-9

18. Gedik, B., Schneider, S., Hirzel, M., Wu, K.: Elastic scaling for data stream processing. IEEE Trans. Parallel Distrib. Syst. **25**(6), 1447–1463 (2014). https://doi.org/10.1109/TPDS.2013.295
19. Golab, L., Garg, S., Özsu, M.T.: On indexing sliding windows over online data streams. In: Bertino, E., et al. (eds.) EDBT 2004. LNCS, vol. 2992, pp. 712–729. Springer, Heidelberg (2004). https://doi.org/10.1007/978-3-540-24741-8_41
20. Google Cloud Dataflow. https://cloud.google.com/dataflow/
21. Heinze, T., Pappalardo, V., Jerzak, Z., Fetzer, C.: Auto-scaling techniques for elastic data stream processing. In: Bellur, U., Kothari, R. (eds.) The 8th ACM International Conference on Distributed Event-Based Systems, DEBS 2014, Mumbai, India, 26–29 May 2014, pp. 318–321. ACM (2014). http://doi.acm.org/10.1145/2611286.2611314
22. Hirzel, M., Soulé, R., Schneider, S., Gedik, B., Grimm, R.: A catalog of stream processing optimizations. ACM Comput. Surv. **46**(4), 46:1–46:34 (2013). http://doi.acm.org/10.1145/2528412
23. Kang, J., Naughton, J.F., Viglas, S.: Evaluating window joins over unbounded streams. In: Dayal, U., Ramamritham, K., Vijayaraman, T.M. (eds.) Proceedings of the 19th International Conference on Data Engineering, 5–8 March 2003, Bangalore, India, pp. 341–352. IEEE Computer Society (2003). https://doi.org/10.1109/ICDE.2003.1260804
24. Kombi, R.K., Lumineau, N., Lamarre, P.: A preventive auto-parallelization approach for elastic stream processing. In: Lee, K., Liu, L. (eds.) 37th IEEE International Conference on Distributed Computing Systems, ICDCS 2017, Atlanta, GA, USA, 5–8 June 2017, pp. 1532–1542. IEEE Computer Society (2017). https://doi.org/10.1109/ICDCS.2017.253
25. Mukhopadhyay, A., Mazumdar, R.R.: Analysis of randomized join-the-shortest-queue (JSQ) schemes in large heterogeneous processor-sharing systems. IEEE Trans. Control Netw. Syst. **3**(2), 116–126 (2016). https://doi.org/10.1109/TCNS.2015.2428331
26. Nasir, M.A.U., Morales, G.D.F., García-Soriano, D., Kourtellis, N., Serafini, M.: The power of both choices: practical load balancing for distributed stream processing engines. In: Gehrke, J., Lehner, W., Shim, K., Cha, S.K., Lohman, G.M. (eds.) 31st IEEE International Conference on Data Engineering, ICDE 2015, Seoul, South Korea, 13–17 April 2015, pp. 137–148. IEEE Computer Society (2015). https://doi.org/10.1109/ICDE.2015.7113279
27. Neumeyer, L., Robbins, B., Nair, A., Kesari, A.: S4: distributed stream computing platform. In: Fan, W., et al. (eds.) ICDMW 2010, The 10th IEEE International Conference on Data Mining Workshops, Sydney, Australia, 13 December 2010, pp. 170–177. IEEE Computer Society (2010). https://doi.org/10.1109/ICDMW.2010.172
28. Peng, B., Hosseini, M., Hong, Z., Farivar, R., Campbell, R.H.: R-storm: resource-aware scheduling in storm. In: Lea, R., Gopalakrishnan, S., Tilevich, E., Murphy, A.L., Blackstock, M. (eds.) Proceedings of the 16th Annual Middleware Conference, Vancouver, BC, Canada, 07–11 December 2015, pp. 149–161. ACM (2015). http://doi.acm.org/10.1145/2814576.2814808
29. Rivetti, N., Anceaume, E., Busnel, Y., Querzoni, L., Sericola, B.: Proactive online scheduling for shuffle grouping in distributed stream processing systems. In: Proceedings of the 17th ACM/IFIP/USENIX International Middleware Conference, Middleware (2016)

30. Rivetti, N., Busnel, Y., Querzoni, L.: Load-aware shedding in stream processing systems. In: Gal, A., Weidlich, M., Kalogeraki, V., Venkasubramanian, N. (eds.) Proceedings of the 10th ACM International Conference on Distributed and Event-based Systems, DEBS 2016, Irvine, CA, USA, 20–24 June 2016, pp. 61–68. ACM (2016). http://doi.acm.org/10.1145/2933267.2933311

31. Rivetti, N., Querzoni, L., Anceaume, E., Busnel, Y., Sericola, B.: Efficient key grouping for near-optimal load balancing in stream processing systems. In: Eliassen, F., Vitenberg, R. (eds.) Proceedings of the 9th ACM International Conference on Distributed Event-Based Systems, DEBS 2015, Oslo, Norway, 29 June–3 July 2015, pp. 80–91. ACM (2015). http://doi.acm.org/10.1145/2675743.2771827

32. Sattler, K., Beier, F.: Towards elastic stream processing: patterns and infrastructure. In: Cormode, G., Yi, K., Deligiannakis, A., Garofalakis, M.N. (eds.) Proceedings of the First International Workshop on Big Dynamic Distributed Data, CEUR Workshop Proceedings, Riva del Garda, Italy, 30 August 2013, vol. 1018, pp. 49–54. CEUR-WS.org (2013). http://ceur-ws.org/Vol-1018/paper9.pdf

33. Schneider, S., Andrade, H., Gedik, B., Biem, A., Wu, K.: Elastic scaling of data parallel operators in stream processing. In: 23rd IEEE International Symposium on Parallel and Distributed Processing, IPDPS 2009, Rome, Italy, 23–29 May 2009, pp. 1–12. IEEE (2009). https://doi.org/10.1109/IPDPS.2009.5161036

34. Senderovich, A., Weidlich, M., Gal, A., Mandelbaum, A.: Queue mining – predicting delays in service processes. In: Jarke, M., et al. (eds.) CAiSE 2014. LNCS, vol. 8484, pp. 42–57. Springer, Cham (2014). https://doi.org/10.1007/978-3-319-07881-6_4

35. Stonebraker, M., Çetintemel, U., Zdonik, S.B.: The 8 requirements of real-time stream processing. SIGMOD Rec. **34**(4), 42–47 (2005). https://doi.org/10.1145/1107499.1107504

36. Sullivan, M., Heybey, A.: Tribeca: a system for managing large databases of network traffic. In: Douglis, F. (ed.) 1998 USENIX Annual Technical Conference, New Orleans, Louisiana, USA, 15–19 June 1998. USENIX Association (1998). https://www.usenix.org/conference/1998-usenix-annual-technical-conference/tribeca-system-managing-large-databases-network

37. Vengerov, D., Menck, A.C., Zaït, M., Chakkappen, S.: Join size estimation subject to filter conditions. PVLDB **8**(12), 1530–1541 (2015). http://www.vldb.org/pvldb/vol8/p1530-vengerov.pdf

38. Wu, Y., Tan, K.: ChronoStream: elastic stateful stream computation in the cloud. In: 2015 IEEE 31st International Conference on Data Engineering, pp. 723–734, April 2015. https://doi.org/10.1109/ICDE.2015.7113328

39. Xu, J., Chen, Z., Tang, J., Su, S.: T-storm: traffic-aware online scheduling in storm. In: IEEE 34th International Conference on Distributed Computing Systems, ICDCS 2014, Madrid, Spain, 30 June–3 July 2014, pp. 535–544. IEEE Computer Society (2014). https://doi.org/10.1109/ICDCS.2014.61

40. Xu, L., Peng, B., Gupta, I.: Stela: enabling stream processing systems to scale-in and scale-out on-demand. In: 2016 IEEE International Conference on Cloud Engineering, IC2E 2016, Berlin, Germany, 4–8 April 2016, pp. 22–31. IEEE Computer Society (2016). https://doi.org/10.1109/IC2E.2016.38

41. Zaharia, M., Das, T., Li, H., Hunter, T., Shenker, S., Stoica, I.: Discretized streams: a fault-tolerant model for scalable stream processing. Technical report UCB/EECS-2012-259. Department of Electrical Engineering and Computer Science, California University, Berkeley (2012). http://www.eecs.berkeley.edu/Pubs/TechRpts/2012/EECS-2012-259.html

SSG: An Ontology-Based Information Model for Smart Grids

Khouloud Salameh[1], Richard Chbeir[2]([⊠]), and Haritza Camblong[3]

[1] American University of Ras Al Khaimah, Ras Al Khaimah, UAE
[2] Univ. Pau & Adour Countries, E2S-UPPA, LIUPPA, EA3000,
64600 Anglet, France
rchbeir@acm.org
[3] University of the Basque Country, 20018 Donostia, Spain

Abstract. Nowadays, an electricity blackout can have a domino effect on the overall power system, causing extremely bad effects on the economical, ecological and operational countries perspectives. All this emphasizes the need for conceiving an upgraded vision of today's and tomorrow's power systems that have to be smart to meet the society expectations. Smart grids have been emerging as an appropriate solution for such needs. This work addresses two main related challenges encountered in the management of such power systems: (1) the semantic interoperability needed between their heterogeneous components in order to ensure seamless communication and integration, and (2) a means to consider their various objectives from economical, ecological, and operational perspectives, to mention some. In this paper, we propose a three-layered smart grid management framework, aiming at resolving these two issues. The backbone of the framework is *SSG*, a generic ontology-based model, detailed here. It aims at modeling the smart grid components, their features and properties, allowing the achievement of the smart grid objectives. Several evaluations have been conducted in order to validate our proposed framework and emphasize the *SSG* importance and utility in the energy domain. Obtained results are satisfactory and draw several promising perspectives.

Keywords: Information modeling · Ontology · Power system Smart grid

1 Introduction

In the era of new technologies and with the growing need for reliable ecological energy supplies [8], current electrical grids have to be upgraded in order to be smarter, more flexible and able to operate, monitor and heal themselves autonomously. Here comes the *SG* as one of the main contributor in the power systems update. However, there are several challenges have to be solved before. One of the most important challenges is related to heterogeneity. In essence, *SGs* consists of a number of heterogeneous components (built and supplied by

© Springer-Verlag GmbH Germany, part of Springer Nature 2019
A. Hameurlain et al. (Eds.): TLDKS XL, LNCS 11360, pp. 94–124, 2019.
https://doi.org/10.1007/978-3-662-58664-8_4

different companies, for different purposes, and using various protocols [16]). In addition, the heterogeneity of such power systems would arise further from the internal and external interactions of their components as well as with the external environment. All this underlines the need of an appropriate semantic interoperability ensuring a seamless information exchange between components within three layers as discussed in [14,15]: Field Layer, Knowledge Layer, and Management Layer. The three layers will be briefly described in what follows.

- **Field Layer (FL):** Via this layer, the data collector gathers all data exchanged between components via a low-level communication environment relying on standardized protocols (e.g., BACnet, Modbus, etc.). Once gathered, those data are stored in a low-level data repository and pushed up to the next layers.
- **Knowledge Layer (KL):** In order to resolve the interoperability issues and open up the possibility to model the new trends in today's energy systems (i.e., prosumers, electric vehicle, etc.), it is essential to capture and understand the semantics of exchanged data to ensure a seamless communication between the components within the power system. Through this layer, the semantic middleware insures the semantic translation of the collected data using our proposed ontology-based information model called *SSG*. Furthermore, the reasoner is responsible of processing information and using it to infer additional value thanks to many rules and constraints defined in this layer.
- **Management Layer (ML):** In this layer, a collaborative diagnostics, a self-optimization for disturbance, and a remote visualization for the users (via an integrated simulation and synthesis) are provided. Besides, the information extracted from the knowledge layer is processed in order to achieve the objectives of the power system. To do so, a battery of advanced management services (e.g., Demand side management, minimization of transmission losses, etc.) is designed.

In addition to the operational aspect related to the components operating, the *SG* needs to ensure several services each targeting a different objective. First, a *SG* aims at providing reliable and secured identification when incorporating heterogeneous components. In today's digital world, cyber-attacks [18,20], such as intentionally switching off the *SG* operators, could cause cascade damages on the grid. Hence, it is important to provide such an identification for the components helping in reducing the grid intrusions. Second, each component can play multiple roles, participating in the emergence of a new paradigm known as 'Prosumer' [12,21], referring to the components able to **PRO**duce power and con**SUME** energy at the same time. Hence, an *SG* can be seen as a multi-objective system that depends on a potential interaction among different stakeholders (i.e., energy sources, energy consumption loads, etc.), having each its objectives, which emphasizes the need of taking into account all the aspects involved in the achievement of the *SG* objectives. Third, the *SG* needs to cope with the mobility of the several components (e.g., electric vehicles, boats, etc.) during their lifetime. Fourth, an *SG* would become an important player in the electricity market relying on its components participation in the environment.

The goal of this study is to address the above issues and challenges by providing an appropriate information modeling for *SGs*. In other words, our goal is to propose a recommended data model for *SG* description, allowing to create an interoperable power system that enables the integration and the validation of the various new heterogeneous renewable distributed generation systems and various storage technologies.

In this paper, we present a dedicated framework for better management of *SG* driven by adapted tools and services. We also detail here our ontology-based *SG* model called *SSG*, capable of: (1) being compliant and aligned with existing information models, coping with the interoperability between all the layers, providing the reasoning capabilities and smart features needed, as well as (4) solving the multi-objective aspect of the *SG*.

The rest of this paper is organized as follows. Section 2 presents the state of the art of existing power systems information models. Section 3 presents our *SSG* ontology through its main concepts. Section 4 describes the evaluation methodology and results of the proposed framework and ontology. Section 5 concludes the paper.

2   Related Work

Several approaches have been provided in the literature addressing the problem of 'Power system information modeling'. They can be categorized into syntactic-based and semantic-based approaches. The syntactic-based models are intended to provide a standard way to represent the data of the system. The semantic-based models are ontology-based information models, aiming at providing a richer and complex knowledge representation about the entities and relations between them.

2.1   Syntactic Based Models

2.1.1   Common Information Model
The Common Information Model (CIM) [19] is a widely accepted electricity information model being part of the IEC 61970 standards. Its main objective is to develop a platform independent data model for enabling better grid interoperability. This model includes the exchange between market participants and market operators as well as communication between market operators. In the CIM model, the *PowerSystemResource* concept is composed of the *Equipment* concept that contains the components of a power system that are physical devices, electronic or mechanical. Two types of equipment exist: (1) *ConductingEquipment* and (2) *Powertransformer*. A *ConductingEquipment* concept, represents the parts of the power system that are designed to carry current. A *Powertransformer* is an electrical device, allowing a mutual coupling between electric circuits.

From the multi-objective perspective, the CIM model [19] does not fully describe all the operational properties of the distributed energy sources and the storage systems. In addition, it covers partially the ecological aspect (using the *EmissionType* parameter) and the economical aspect (using the *CostPerEnergyUnit* and *CostPerHour* parameters). The identification aspect is limited to only two parameters: *Id*, *Name*. However, the mobility and the multi-role aspects were totally absent in the model. From the interoperability perspective, the CIM model does not cover completely the field layer. In addition, since it is an UML based model, this impoverishes the semantic relations between the concepts, which limits its knowledge coverage. In addition, as mentioned before, since there is a lack in representing all the objective aspects of a power system, this also affects negatively the management layer.

2.1.2   MIRABEL FlexEnergy Data Model

The MIRABEL smart grid system [23] comes to hand over the flexibility in energy demand and supply. It incorporates the power profile concept which associates a consumption/production schedule for each branch.

In order to achieve such flexibility in energy demand and supply in the power grid, a data model has been developed in [23] consisting of five main classes: *branch, actor, energyprofile, constraint* and *flex-offer*. A *branch* is an energy consumer or producer that has a specific energy load over a certain time span (called *energyprofile*). An *actor* has minimum or maximum demands (called *constraints*) on their energy load, price and time. These constraints are issued (by an actor) toward the branches owned by the actor. The *flex-offer* class defines two types of demands: flexible demand and non-flexible demand. Flexible demand can often be shifted from the peak demand times to lower demand times, while non-flexible demand should be satisfied immediately.

From the multi-objective perspective, the model in [23] provides a high economical aspect representation and a slighter representation of the operational and identification aspects, since it is dedicated to conceive a flexible market power exchange. However, the ecological, mobility and multi-roles aspects are absent in it. From the interoperability perspective, the MIRABEL model does not cover completely the field layer. Similarly to the CIM model, Mirabel is an UML based model, which impoverishes its semantic expressiveness and the knowledge coverage. In addition, as mentioned before, since there is a lack in representing all the objective aspects, affecting negatively the management layer.

2.1.3   Facility Smart Grid Information Model

The Facility Smart Grid Information Model (FSGIM) [7] is developed with the aim of enabling energy consuming branches and control systems in the customer premises so to manage electrical loads and energy sources in response to communications with the smart grid. To achieve this, an object-oriented information model is defined to support a wide range of energy management applications and electrical service provider interactions. The proposed information model [7] provides a common basis to describe, manage, and communicate information on aggregate electrical energy consumption and forecasts.

From the multi-objective perspective, the FSGIM model covers almost all the components of a power system, except the storage devices (only the thermal storage systems are modeled). However, the model takes fully into account the economical and identification aspects. Concerning the ecological aspect, it is partially covered in the model (using *Emission* parameter). The multi-role aspect is completely absent in the model. From the interoperability perspective, the FSGIM model does not cover completely the field layer. In addition, since it is an object-oriented model, it has a limited means to express the semantic relations between the components and the reasoning capabilities of the system. All this causes a partial management layer coverage.

2.1.4   OASIS Energy Interoperation

OASIS Energy Interoperation [4] enables collaborative use of energy in a power network. It defines XML-based vocabularies for the interoperable and standard exchange of information related to energy prices and bids (demand and response), network reliability, emergency signals and the prediction of loads consumption. This information relies on the $WS - Calendar$ [5] and $EMIX$ (electricity market Information Exchange Specification) [3]. The first defines how to specify and communicate the duration and time of a schedule, while the later specifies the semantics in electricity markets.

From the multi-objective perspective, the OASIS model covers completely the economic aspect since it targets the electricity market information model. However, it neglects the remaining aspects. From the interoperability perspective, the OASIS model covers partially the three layers, since it does not cover completely all the components and operational parameters, without taking into account all the semantic relations between the components.

2.2   Semantic Approaches

2.2.1   Facility Ontology

The Facility Ontology[1] aims at conceiving a standard nomenclature for the power systems, by providing a representation of its components and their control parameters. Complying with the Suggested Upper Merged Ontology (SUMO), the proposed ontology aims to classify the power system in two main concepts: the *Physical* and the *Abstract* concepts. The *Physical* concept serves for describing the physical components of the power system (i.e., production unit, storage unit, consumption unit and conversion unit) with a set of related properties. Concerning the *Abstract* concept, two concepts are introduced: the *Management* concept, and the *Policy* concept. The *Management* concept consists of four sub concepts: (i) the *Energy_trading*, (ii) the *Lc_operation*, (iii) the *Mgcc_operation* and (iv) the *Operational_modes*. The *Lc_operation* and *Mgcc_operation* concepts contain all the information related to the load and central controllers. The *Energy_trading* concept represents the information related

---

[1] https://github.com/usnistgov/facility.

to the power exchanged in the grid, such as the power prices, the minimum and the maximum power quantity. And finally, the *Policy* concept, refers to the information related to the constitution (*Design* concept), the operation (*Operation* class) and interface (*integration* concept) of the power system.

From the multi-objective perspective, the ontology shows a high efficiency in representing the operational aspect, by modeling all the components of the power system. Similarly to the operational aspect, the economical one was taken into account via the *Energy_trading* concept. The identification aspect was limited to the definition of the *ID*, *Mode* and *Manufacturer* parameters. However, the mobility, the ecological and the multi-role aspects were totally absent in the ontology. From the interoperability perspective, the Facility Ontology covers completely the field layer. However, it is poor in representing the semantic relations between the components (limited to the "hasSubClass" relations), which limits its knowledge coverage. In addition, as mentioned before, there is a lack in representing all the objective aspects of a power system which affects negatively the management layer.

2.2.2   Prosumer Ontology

In [9], the authors propose a classification of the power system components using several predefined scenarios. Based on the UK property classification, five power consumption patterns are identified, namely: (1) *commercial premises* consisting of the consumers having varying operating times, (2) *business related premises* consisting of the consumers having fixed operating times (e.g., office times), (3) *residential premises* consisting of the houses consumption, (4) *non − residential premises* consisting of non-residential premises (e.g., hospitals, schools, etc.) having more critical power needs, and (5) *industrial premises* consisting of the factories consumption having uninterrupted power needs. Concerning the energy sources classification, two categories were also introduced in [9]: *renewable* and *non − renewable* energy sources, while three energy storage systems categories were identified, according to the type, produced power and charge and discharge efficiency, namely: (1) *energy management*, (2) *power quality*, and (3) *bridging power*. In addition, the *component connectivity* focuses on enabling the exact connectivity relationships between the producers and the consumers. And finally, the *Service Contracts* comes to describe the information exchanged between the producers and the consumers in a competitive market. It contains the Start/End Date of the contract, the type of payment and the charges per units of power.

From the multi-objective perspective, the ontology in [9] shows a lack in the operational aspect, since it is limited to modeling the main components of a power system, without taking into account their operational parameters. When it comes to the economic aspect, it is partially taken into account by modeling the contracts between producers and the consumers. The ecological aspect is partially modeled by distinguishing the renewable and non-renewable energy sources. The remaining aspects are totally absent in this model [9]. From the interoperability perspective, the Prosumer ontology covers partially the field layer. This affects directly the knowledge layer modeling. Here again, the management layer can partially be addressed due to the lacks in the multi-objective aspect modeling.

2.2.3 Upper Ontology for Power Engineering Application

Based on the Common Information Model (CIM) [19], the authors in [2] propose an ontology that mainly aims at monitoring the health status of the power systems. In this model, the concept *Measurement* represents anything that can be measured, including data taken from sensors and historical data. In addition, anything that is extracted from raw data is represented as an *Interpreted Data*, and specifically as a *Summary Interpretation* or a *Detailed Interpretation*. Moreover, the components' operations in the system are represented via the *Agent Action*. This model supports the exchange of messages between agents, but not explicitly defined. Although adopted by several applications, the upper ontology usually needs to be enriched with additional concepts to cover all the required information.

From the multi-objective perspective, and since this model [2] is based on the CIM [19], this leads to inherit the same objective aspects coverage. Hence, the upper ontology covers partially the operational, identification, economical and ecological aspects, but doesn't take into account the mobility and multi-roles aspects. From the interoperability perspective, the upper ontology covers partially the field layer. In addition, it neglects the semantic relations between the components, which makes the knowledge layer incomplete. All this causes a lack in the management layer.

2.3 Summary

In this section, we present a comparison summary between the existing approaches, highlighting their strengths and drawbacks with respect to their ability to resolve the interoperability issue within a power system, and the integration of the necessary aspects allowing the achievement of related services. Three symbols for comparison will be used in whats follows: (1) '−' to express the low capabilities of an approach in covering a feature, (2) 'partial' when an approach has middle coverage capabilities, and (3) '+' to express the high coverage capabilities of an approach.

2.3.1 Interoperability Aspect

Table 1 shows the ability of the existing approaches to cope with the interoperability issue. In short, most of them cover the modeling of the field layer, which contains the physical components of the power systems. Concerning the Knowledge/Information layer, the semantic-based approaches show a better potential in the knowledge modeling, compared to the syntactic-based ones, represented by the classification and the categorizing of the power systems components, but lack in fully modeling the relationships between them. Table 1 also shows that existing approaches cannot provide an appropriate modeling of the management layer, since they are mostly limited to modeling the electricity market information.

Table 1. Comparison of existing power system information models with respect to the interoperability aspect

	Interoperability layers		
	Field layer	Knowledge/Information layer	Management layer
CIM [19]	Partial	Partial	Partial
FSGIM [7]	+	Partial	Partial
OASIS [4]	−	−	−
MIRABEL [23]	−	−	−
Prosumer [9]	Partial	Partial	Partial
Facility ontology (see footnote 1)	+	Partial	Partial
EFEFEFUpper ontology [2]	Partial	Partial	Partial

2.3.2   Multi-objective Aspect

Table 2 summarizes the main commonalities and differences between existing approaches with respect to the six categories of aspects used in the achievement of the Power Systems objectives. In short, few take properly into account the identification aspect. In contrast, the operational aspect is the core of most of the existing models, whose aim was to standardize the technical vocabulary in the power systems, except MIRABEL system which mainly focuses on the electricity market modeling. Clearly, as the comparison table shows, the economical aspect is highly modeled since most of the existing models aim at conceiving a market power exchange. Moreover, the ecological aspect is merely modeled through a small set of properties related to the gas emission of the components. However, two aspects are almost absent in the existing information models, namely: (1) the mobility aspect representing the shifts of the components in the system, and (2) the multi-roles aspect, representing the roles played by a component during its lifetime according to a certain context. To sum up, none of the existing approaches completely addresses the interoperability and the mutli-objective aspect of the power system. In the following section, we provide our *SG* Management System framework, aiming at resolving interoperability issues from the information perspective by integrating all the power system aspects related to its objectives.

Table 2. Comparing existing power system information models regarding the *SG* multi-aspect

	MG objective aspect					
	Identification	Operation	Mobility	Economy	Ecology	Multi-roles
CIM [19]	Partial	Partial	−	Partial	Partial	−
FCGIM [7]	+	Partial	Partial	Partial	+	Partial
OASIS [4]	−	−	−	+	−	−
MIRABEL [23]	Partial	Partial	−	+	−	−
Prosumer [9]	−	Partial	−	−	Partial	−
Facility ontology (see footnote 1)	Partial	+	−	+	−	−
Upper ontology [2]	Partial	Partial	−	Partial	Partial	−

3 *SSG* Ontology

As seen in our related work study, semantic-based models showed a higher expressive power in dealing with interoperability issues and to some extend with the multi-objective aspect of the *SGs*. Thus, this drove us to adopt a semantic-based approach called *SSG*, a generic ontology-based model, aiming at modeling the *SG* components, their parameters and additional properties allowing the achievement of its objectives.

3.1 Why "Ontologies are Appropriate" Means for Semantic Approaches?

Due to its importance [13] in information systems and artificial intelligence, an ontology-based *SG* information model would provide a shared knowledge conceptualization allowing an easier system interaction and manipulation, especially for non-computer scientists, while giving the grid reasoning capabilities and autonomy.

3.1.1 Ontology as a Shared Knowledge

Since an *SG* consists of a number of heterogeneous components, it is important to define a shared representation of the exchanged information. In addition, each component has a direct/indirect impact on the other components and on the overall grid.

3.1.2 Ontology as a Better Means for Information Retrieval

Since a power system is usually managed by non-computer-scientists, an ontology would help them interact and manipulate the system in an easier and more intuitive way. Besides, an ontology would provide a structure that is flexible, and that naturally organizes the information in multidimensional ways.

3.1.3 Ontology as a Reasoning Strategy

Due to the intermittent aspect [6] of the renewable energy sources and the exposure of the power system to predictable and non-predictable events (power system anomalies, storms, etc.), an ontology becomes essential since it can also represent beliefs, goals, hypotheses, and predictions. These latter will give the components the ability to act and react autonomously or collectively according to a certain event or goal.

3.2 *SSG* Overview

While conceiving an ontology, the main target is to settle a shared terminology describing the power system. Several steps were conducted while developing our ontology [22]. In the aim of being compliant with existing standards, the first step was to identify the well-known and most adopted standards in the power domain.

Two important standards have been identified: the CIM/IEC 61790 model, and the IEC 61850-7-420 related to the basic communication structure for distributed energy resources logical nodes. The second step consisted of grouping the concepts into categories in order to check the coverage of the ontology regarding the needed aspects. And finally, the refinement phase consisted of establishing the semantic relations between the defined concepts. Thus, to cope with the interoperability issues, the skeleton structure of the *SG* (called the **basic structure**) is mainly based on the CIM standard and the multi-objective aspect (called **extended structure**) is based on the IEC 61850 standard and completed with a set of additional properties.

3.3 Why CIM and IEC 61850

CIM is an open standard for representing power system components developed by the Electric Power Research Institute (EPRI) in North America. The standard was developed as part of the IEC TC57 WG13 on developing a Control Centre Application Programming Interface (CCAPI) to provide a common model for describing the components in power systems for use in a common Energy Management System (EMS) Application Programming Interface (API). Besides the fact that the CIM is a standardized data model, this format has been adopted by the major EMS vendors to allow the exchange of data between their applications, independent of their internal software architecture or operating Platform.

IEC 61850 is a standardized data model for representing distributed energy resources (DER), which comprise dispersed generation devices and dispersed storage devices, including reciprocating engines, fuel cells, microturbines, photovoltaics, combined heat and power, and energy storage. The IEC 61850 is now an International Standard, that addresses most of the issues that migration to the digital world entails, especially, standardization of data names, creation of a comprehensive set of services, implementation over standard protocols and hardware, and definition of a process bus. Multi-vendor interoperability has been demonstrated and compliance certification processes are being established.

All the aforementioned reasons mentioned above, lead us to adopt both standards in the aim of being compliant with international norms and protocols. Our ontology, called *SSG*, is a graph representing a collection of subject-relation-object triples, where:

- Nodes designate subjects, objects, or subject/object properties: (1) *SG* branches and components (e.g., EnergyStorageBranch, WindTurbine, etc.), and (2) Corresponding property values (e.g., panelWidth, totalCost, etc.)
- Edges connecting source/destination nodes, designate relations: (1) Relations between components (e.g., WindTurbine isA DistributedEnergySource, etc.), and (2) Property and value relations (e.g., windTurbine HasSpeed 50, solarPanel HasCost 7500, etc.)

The property values and edges in *SSG* are mainly classified into five categories: identification, mobility, operation, economic, and ecology. Details are provided in what follows.

3.4 *SSG* Basic Structure

To cope with the interoperability issues, our *SSG* basic structure is a semantic translation of the CIM extension proposed in [24]. Knowing that the CIM is not dedicated to cover specifically the *SG* components modeling, the authors in [24] proposed additional features (e.g., solar power, wind power, etc.). Here comes the importance of our ontology that represents in a simple and clean way, each branch structure which contains the set of the equipment that composes it. Figure 1 shows the 'Microgrid' concept, inheriting from the 'CIM:SubControlArea', which describes relative information of the power system operation and allows the creation of several connected power systems instances. Based on the branch concept defined in [24], four main branches are added here: (1) Distributed energy resource branch, (2) Energy storage branch, (3) Electrical load branch, and (4) Infrastructure Branch, where each has its own Branch Switch and Branch Controller. The Branch Switch is responsible of turning on/off the branch, and the Branch controller is the manager of the branch operations. All concepts borrowed from CIM have been prefixed with 'CIM:' in the following figures of the provided ontology.

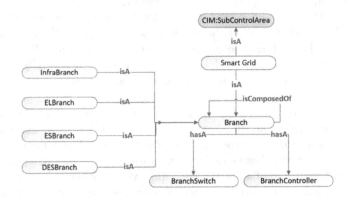

Fig. 1. Extract of the *SSG* skeleton structure

3.4.1 Distributed Energy Resource (DER) Branch

The distributed energy resource branch consists of renewable or non-renewable energy sources. Figure 2 shows the DER branch concept, consisting of a Solar Power Branch, Wind Power Branch, Combined Heat Power Branch and Fuel Power Branch.

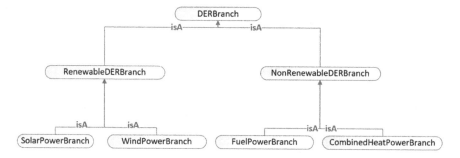

Fig. 2. Extract of the DER branch

Note that a branch is a combination of several equipment, when working together, they accomplish a specific function in the *SG* (e.g., a Solar Cell and a Converter are two main equipment constituting the Solar Power branch and allowing its functioning in the power system). In more details, a Solar Power branch (cf. Fig. 3) consists mainly of a Solar Cell and a converter. The Solar Cell is an electrical device that converts the energy of light directly into electricity by the photovoltaic effect, which is a physical and chemical phenomenon. The converter is a branch for altering the nature of an electric current or signal, especially from AC to DC (Ac/Dc Converter) or vice versa (commonly called Inverter). This latter can be a Monophasic inverter or a Triphasic inverter.

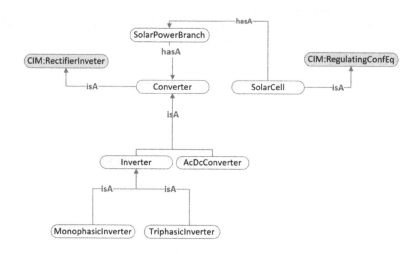

Fig. 3. Extract of the photovoltaic branch package

Figure 4 depicts the wind power branch. It includes mainly, the wind turbine and the converter. The wind turbine generates electricity from the kinectic power of the wind. The wind turns two or three propeller-like blades around a rotor. The rotor is connected to the main shaft, which spins a generator to create electricity. Similarly to the photovoltaic branch, the converter consists an essential component in the wind power structure.

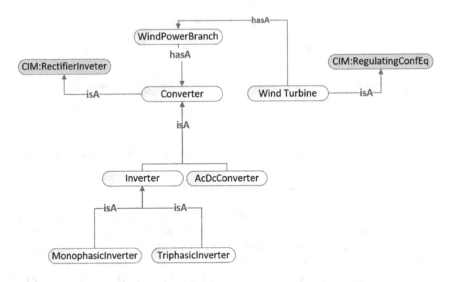

Fig. 4. Extract of the wind power branch

3.4.2   Energy Storage (ES) Branch

Recently, the energy storage systems start to have a great potential in radically transforming the global energy landscape, helping to solve key issues in the integration of renewable energy systems. Energy storage systems play an essential role in stabilizing the SG, improving the quality of power supply, and achieving power peak shaving. The energy storage branch consists mainly of the energy storage device (e.g., Pumped-Storage Hydroelectricity (PSH), batteries, etc.) and a converter (cf. Fig. 5).

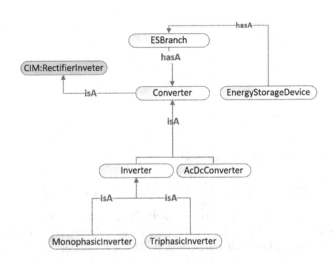

Fig. 5. Extract of the energy storage branch

3.4.3 Electrical Load (EL) Branch

An electrical Load is an electrical component or branch that consumes electric power. It is mainly consisting of the electrical appliance components (cf. Fig. 6).

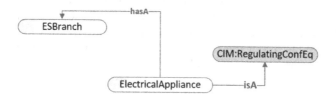

Fig. 6. Extract of the electrical load branch

3.5 *SSG* Extended Structure

To cope with the multi-objective aspect of an *SG*, *OntoMG* aims to model all the aspects/functionalities participating in the achievement of its objectives. Hence, six concepts are defined, each covering an objective aspect, namely: (1) identification, (2) economical, (3) operation, (4) mobility, (5) ecological and (6) multi-roles. Those concepts are the key for conceiving an *SG* able to reason and act autonomously.

3.5.1 Identification Concept

An *SG* consists of several heterogeneous branches, each having its own characteristics and operation modes during its lifetime. Thus, when joining an *SG*, each branch is associated, through an identification service, with an 'identity' consisting of a number of properties distinguishing it from the others and giving it the possibility to be automatically recognized. The identification concept consists of a number of properties (cf. Table 3): the serial number which is a unique value, the type, brand and model designating a certain provider.

Table 3. Identification concept

Name	Description	Type
Serial#	Unique identifier of a component within the system	String
Type	Type to which a component belongs	String
Brand	Feature that distinguishes one seller's component from those of others	String
Model	Style or design of a particular component	String

3.5.2 Economic Concept

Due to the importance of the *SG* from economic perspective, it is essential to consider related properties of its components. Those properties imply several features related to the *SG* participation in the electricity market. Table 4 shows the main properties of the economic aspect consisting of: the maintenance cost, the total cost, the start up cost, the stop cost, the installation cost, the equipment cost and the operating cost. Two additional properties are only assigned to the branches being able to sell their produced/stored power are the power price per KWh, the power price per hour, and the power cost.

Table 4. Economic concept

Name	*Description*	*Type*
EqCost	Equipment cost of a component	Number
MaintenanceCost	Maintenance cost of a component	Number
InstallCost	Installation cost of a component	Number
OpCost	Operating cost of a component	Number
TotalCost	Total cost of a component	Number
StrCost	Start up cost of a component	Number
StopCost	Stop cost of a component	Number
PwrKWhPrice	Power price vector per KWh	Number
PwrhPrice	Power price vector per hour	Number
PwrCost	Production power cost vector	Number
CptBill	Consumption bill vector	Number

3.5.3 Operation Concept

The operation concept encompasses the technical properties related to the components functioning during their lifetime in the power system. Since our model is based on the IEC 61850 in its extended structure, this eases the exchanges of the technical information between the *SG* components.

Tables 5, 6 and 7 show the list of the distributed energy source (DES), energy storage (ES) and electrical load (EL) operation properties, respectively.

Table 5. DER operation concept

Name	Description	Type
$IEC : VRtg$	Voltage level rating	Number
$IEC : ARtg$	Current rating under nominal voltage under nominal power factor	Number
$IEC : HzRtg$	Nominal frequency	Number
$IEC : TmpRtg$	Max temperature rating	Number
$IEC : VARtg$	Max volt-amps rating	Number
$IEC : WRtg$	Max watt rating	Number
$IEC : Vartg$	Max var rating	Number
$IEC : MaxWOut$	Max watt output - continuous	Number
$IEC : WRtg$	Rated Watts	Number
$IEC : MinWOut$	Min watt output - continuous	Number
$IEC : EffRtgPct$	Efficiency at rated capacity as percent	Number
$LaunchCount$	Number of time the components is launched during an interval of time	Number
$Penalty$	Waiting time penalty of launching the component	Number
$SInit$	Desired schedule of the component	Double
SOp	Operational schedule of the component	Double

Table 6. ES operation concept

Name	Description	Type
$IEC : AhrRtg$	Amp-hour capacity rating	Number
$IEC : BatVNom$	Nominal voltage of battery	Number
$IEC : BatSerCnt$	Number of cells in series	Number
$IEC : BatParCnt$	Number of cells in parallel	Number
$IEC : DisChaCnt$	Discharge curve	Number
$IEC : DisChaTim$	Discharge curve by time	Number
$IEC : DisChaRte$	Self discharge rate	Number
$IEC : EffRtgPct$	Efficiency at rated capacity as percent	Number
$IEC : SOCPct$	Battery level as percent	Number
$IEC : SOHPct$	Battery lifetime as percent	Number
$LaunchCount$	Number of time the components is launched during an interval of time	Number
$Penalty$	Waiting time penalty of launching the component	Number
$SInit$	Desired schedule of the component	Double
SOp	Operational schedule of the component	Double

Table 7. EL operation concept

Name	Description	Type
$ActhAm$	A.m active hours	Number
$ActhPm$	P.m active hours	Number
Cpt	Current consumption	Number
$MaxCpt$	Maximum consumption	Number
$MinCpt$	Minimum consumption	Number
$MinStrTim$	Minimum start time consumption	DateTimeStamp
$MaxStrTim$	Maximum start time consumption	DateTimeStamp
$StrTim$	Start time consumption	DateTimeStamp
$MinStopTim$	Minimum stop time consumption	DateTimeStamp
$MaxStopTim$	Maximum stop time consumption	DateTimeStamp
$StopTim$	Stop time consumption	DateTimeStamp
$isPrimary$	Designates a critical load	Boolean
$isSecondary$	Designates a non-critical load	Boolean
$isShiftable$	Designates a shiftable load	Boolean
$LaunchCount$	Number of time the components is launched during an interval of time	Number
$Penalty$	Waiting time penalty of launching the component	Number
$SInit$	Desired schedule of the component	Double
SOp	Operational schedule of the component	Double

3.5.4 Ecology Concept

Knowing the importance of the SG in the integration of green energy production, it becomes essential to take into account the components contribution in the environment. This participation is modeled through ecology concept (cf. Table 8) using several properties, such as the carbon emission ratio, the Ethylene emission ratio, and others gas emissions ratios, expressed in g/Kg. In addition, the pollution costs related to the toxic emissions are modeled using several properties: Carbon Emission Cost, Etyl Emission Cost.

Table 8. Ecology concept

Name	Description	Type
$CarbEss$	Carbon emission ratio	Number
$EthylEss$	Ethyl emission ratio	Number
$HeatEss$	Heat emission ratio	Number
$CarbEssCost$	Carbon emission Cost	Number
$EthylEssCost$	Ethyl emission Cost	Number
$HeatEssCost$	Heat emission Cost	Number

3.5.5 Mobility Concept

In order to model the components ability to move during their lifetime in the SG, a two-dimensional tracking is represented through two concepts: 'Time tracking' and 'Position tracking'. Each concept has a set of properties allowing a fine-grained tracking (cf. Tables 9 and 10).

Table 9. Time tracking concept

Name	Description	Type
DepTim	Departure Time of a mobile component	DateTimeStamp
ArvTim	Arrival Time of a mobile component	DateTimeStamp

Table 10. Location tracking concept

Name	Description	Type
Ctry	Country	String
Lat	Latitude	Double
Long	Longitude	Double
PosInMG	Position in the *SG*	String

3.5.6 Multi-roles Concept

Future *SG* are going through comprehensive changes, especially due to the integration of the Prosumers, where an entity can consume and produce simultaneously in a complete paradigm shift [12].

Fig. 7. Multi-role concept

Figure 7 shows the 'Role' concept defined to model the different roles that a component can play during their lifetime in the grid. Besides, three additional properties are defined (cf. Fig. 11): the 'RoleCondition', the 'RoleStartTime' and the 'Duration'.

3.6 Discussion

Our SSG ontology design and structure highlight its capabilities in resolving the interoperability issues from the three layers:

- Field Interoprability Layer: This is resolved thanks to the use of the CIM and IEC international standards allowing to be compliant with existing standards in the domain. In addition, new concepts are added aiming at covering new technologies and concepts such as electrical vehicle, etc.
- Knowledge/Information Layer: This is resolved thanks to the adoption of an ontology-based model, which allows the semantic modeling of the data.
- Management Layer: This is resolved thanks to the integration of several parameters allowing to cover the six services' categories of the *SGs*:

Table 11. Multi-roles concept

Name	Description	Type
RoleCondition	Required Condition to play a specific role	String
RoleStartTime	Start Time of a specific role	DateTimeStamp
Duration	Play duration of a specific role	Double

– *Identification Services:* the main identification services are the *Authentication* and the *Registration*. In the aim of establishing a secure access to the power system, an *Authentication* service is required. It verifies the identity of any component wishing to access the *SG*. The *Registration* service, is the process of registering the components in the power system using a set of parameters defined in the information/knowledge layer.
– *Operational Services:* the main operational services are: (1) the *Voltage and frequency regulation*, (2) the *Fault detection*, (3) the *Power loss minimization*, and (4) the *Peak power reduction*. The *Voltage and frequency regulation* consists of maintaining a balanced output of the voltage and frequency iof the grid, done despite the systems' disturbances and the load variations. The *Fault detection* consists of detecting power system errors as fast as possible, so that an appropriate action can be immediately taken before major problems can happen. The *Power loss minimization* consists of ensuring the power exchange between the components in a way to reduce the power transmission losses. The *Peak power reduction* consists of reducing the maximum power consumption (for instance, by applying prediction techniques of electrical consumption [11] and demand-side management techniques).
– *Economical Services:* they consist of managing the impact of the components on the electricity market. They play an essential role in delegating the cheapest component that should be launched or implemented to satisfy a certain need. For instance, one main economical service is the *electricity market management* which consists of establishing auction algorithms in order to find the optimal power prices and to maximize the net benefit of the components.
– *Ecological Services:* they consist of managing the participation of the components in the environment. The main ecological service is the *Green decisions management*. It consists of ensuring a cooperation in the power system by gathering the components that have mutual benefits, in order to make green decisions (e.g., putting up consumers having high power needs with the renewable energy sources in the aim of reducing the pollution ratio).
– *Mobility Services:* they are related to the components movements [17] in the power system. The main mobility service is the *Components location tracking*. It consists of determining and tracking the precise location of a component at any time. It is also used by the *Fault detection* service by

facilitating the detection of the location of any problem in order to fix it more rapidly.
- *Multi-roles Services:* they are related to the components which are able to execute many roles during their lifetime in the *SG*. The main multi-roles service is the *Role forcing* which forces a component to play a certain role (i.e., produce, consume or store power) when there is an essential need in the *SG*.

4   Experiments

We conducted several experiments in order to validate our proposed framework and emphasize the *SSG* importance and utility in the electricity domain. Before detailing the conducted tests, it is important to quickly describe the *SSG* design process. We developed *SSG* after exploring the current standards in power domain. In essence, we designed it iteratively by: (1) exploring and comparing the current standards in power domain, (2) presenting our observations and conclusions to several experts, (3) considering their feedback regarding their future needs and expectations. This iterative process has taken almost two years long in order to come up with a stable version. Hence, the feedback and knowledge of the experts have constantly been used to improve the ontology in every iteration.

4.1   Evaluation Criteria

It is worthy to note that there is no unique methodology for developing and evaluating ontologies. Developing ontology is usually an iterative process that can start with a rough first pass at the ontology and then revise and refine the evolving ontology. This process of iterative design will likely continue through the entire lifecycle of the ontology. In our study, we adopted two main quality criteria provided in [10] to evaluate *SSG*:

- **Comprehensibility:** it refers to how easily the language can be understood by technical actors (agents, engineers, etc.). Important aspects are the support of abstraction mechanisms (hiding details), uniform constructs, and a reasonable number of concepts.
- **Domain coverage:** it refers to the ability of the ontology to capture and cover the domain knowledge. It is related to the structure of the provided representation (concepts and relationships) and is the most important aspect of the ontology evaluation.

4.2   Evaluation Context

Although automatic or semi-automatic evaluation techniques are attracting more and more interests, manual evaluation or what is called 'human assessment evaluation' remains commonly adopted in the literature when addressing ontology evaluation [1]. Thus, we conducted manual evaluations to validate the core of *SSG*. We also deployed *SSG* into two projects. Before detailing the obtained results, we detail in what follows: (1) the ontology layers that has been evaluated, (2) corresponding evaluation metrics, and (3) the testers' profiles.

4.2.1   Ontology Layers

Three main ontology layers have been evaluated in our experiments:

- The *syntactic layer* includes respectively the ABox (concepts/classes) and the TBox (instances) of *SSG*
- The *semantic layer* encompasses the semantic relations between concepts (e.g., isA, hasPart, etc.), shaping the structure of the ontology
- The *context layer* includes the additional properties related to the *SG* needs, which are here reflected by its multi-objective aspects.

4.2.2   Evaluation Metrics

In order to correctly evaluate the ontology, three evaluation metrics have been used (the 3Cs requirements [25]):

- The *Correctness* aims at evaluating the clarity of the vocabulary and data of the syntactic layer of the ontology. It is used in our experiments to mainly measure the comprehensibility criteria,
- The *Consistency* targets the evaluation of the semantic layer of an ontology. It is also used to measure the comprehensibility,
- The *Completeness* targets the evaluation of the syntactic and context layers. It aims at evaluating the domain coverage criteria with the services that a *SG* must deal with.

4.2.3   Tests and Testers

Three tests were conducted, each targeting a specific evaluation metric: an ambiguity test, a quiz test, and a real use case scenario to evaluate the correctness, consistency and completeness, respectively. The first two tests were conducted by:

- 80 experts in electrical engineering (45 participants) and electronics (35 participants),
- 45 non-experts in electrical engineering and electronics (mainly computer scientists).

Note that our experts and non-experts are the assistant professors, associate professors, full professors and PhD students of the University of the Basque Country, Spain and the University of Pau and Pays de l'Adour - France.

The choice of having computer scientists in our tests is related to the fact that we believe that future power systems will be multidisciplinary and would require some expertise in Information Technologies in order to understand how things are working together. In what follows, a detailed explanation of each evaluation is presented.

4.3   Comprehensibility Results

In what follows, we show the results obtained with the two metrics of Correctness and Consistency to measure the comprehensibility criteria.

4.3.1   Correctness

A first 'semantic ambiguity test' was done to evaluate the ontology correctness that targets the *syntactic* layer evaluation. A semantic ambiguity refers to the ambiguity of a word to be used in different contexts in order to express different meanings. In this test, the participants were asked to rate the ambiguity degree (if the word is clear/understandable or not) of a list of 60 items on a scale of 0 to 4 (4 expresses a very clear concept with no ambiguity, and 0 expresses a high ambiguity). Those items are categorized into two main categories: the low-level and the high-level items. The low-level items, target the technical data related to the power system structure and branches (i.e., the basic structure). However, the high-level items target the semantic data extracted related to the identification, ecological, economical, operational and mobility concepts (i.e., the extended structure). The obtained results are as follows:

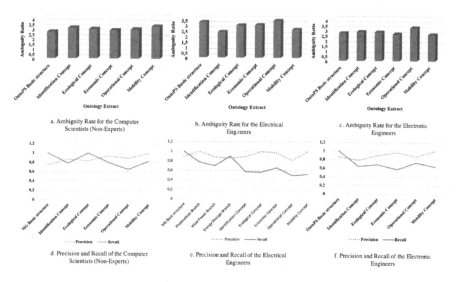

Fig. 8. Experimental results

- For non experts: Figure 8a shows the results of the tests conducted by the 45 testers in computer science. The ambiguity rates vary from 2.66 (Basic structure) to 3.25 (Mobility concept), which can be considered as a very good result for non-experts in the electricity domain. A closer look to the rates led us to conclude that the hardest part was related to the evaluation of the low-level items, driving an ambiguity rate of 2.66. However, it was easier for them to understand the high-level items, resulting an ambiguity rate that varies from 2.85 (Economic concept) to 3.25 (Mobility concept). This is explained by the fact that the computer scientists are less familiar with the technical vocabulary related to the power systems (e.g., solar cell, flywheel, etc.), yet they are globally aware about the high-level concepts related to the electricity market (e.g., Power Price, etc.), ecology (e.g., Gas Emission, etc.), identification (e.g., Serial Number, etc.), and mobility (e.g., Component Position, etc.).

- For electrical engineers: Figure 8b shows the results of the 45 testers in the electrical domain. The ambiguity rates vary from 2.35 (Identification concept) to 3.6 (Operational concept), which is very satisfactory. We observed that the easiest part for electrical experts, contrarily to non-experts, was to evaluate the ambiguity of the technical part, leading to an ambiguity rate of 3.6 (Operation concept). However, it was more difficult for them to understand the high-level items, resulting an ambiguity rate that varies between 2.35 (Identification concept) and 2.975 (Ecology concept).

- For electronic engineers: Figure 8c shows the results of the remaining 35 testers (most of them are students). The ambiguity rates vary between 2.58 (Mobility concept) and 3.25 (Operation concept). A closer look to the rates led us to conclude that the results were not converging, since the lowest ambiguity rate is 2.58 for the mobility concept which is related to the high-level terms, while the highest ambiguity rate is 3.25 for the technical terms. This will allow in the future to measure and compare the **Learning load** of an expert and a non-expert in order to master the proposed vocabulary.

4.3.2   Consistency

A second test was conducted to evaluate the ontology consistency. In this test, the testers were kindly requested to choose the adequate relations between the concepts in a given ontology extract. Similarly to correctness, the list of 6 ontology extracts (each related to an ontology structure and concept) is categorized into two main categories: the low-level and the high-level extracts. The low-level one targets the technical data related to the *SSG* basic structure, while the high-level category targets the semantic data related to the identification, ecology, economic, operation and mobility concepts. For this evaluation, we adopted the precision and recall metrics commonly adopted in Information Retrieval since they meet our needs in evaluating whether the relations between the concepts are relevant or not. Please note that Precision (PR) computes the ratio of the number of correct answers w.r.t. the total number of answers (correct and false), while Recall (R) underlines the number of correctly identified answers w.r.t. the total number of correct answers, including those not answered by the user. The obtained results are as follows:

- For non experts: Figure 8d shows that the highest precision obtained by the computer scientists was reached when dealing with the mobility concept (of 1). This comes from the intuitiveness of the answers (which are the concepts in the ontology such as Country, Latitude and Longitude) that do not need an expertise in the power domain. However, the lowest precision (of 0.74) was reached when dealing with the basic structure. This comes from the specificity of the answers related to the different basic components that compose the 'SG'. On the other hand, Fig. 8d shows that the highest recall (of 1) is reached when dealing with the basic structure. This comes from the fact that since the testers are not experts in the power domain, they chose multiple answers, which increased sometimes the percentage of the correct answers. However, the lowest precision (of 0.658) was reached when dealing with the operation concept. This result confirms our expectation regarding *SSG*.

- For electrical engineers: Figure 8e shows that the highest precision (of 1) obtained by the electrical scientists was reached when dealing with the mobility concept (similarly to the computer scientists). However, the lowest precision (of 0.78) was reached when dealing with the identification concept. This comes from the fact that this concept is brand new for the testers who were assuming that some technical information (e.g., nominal active power, etc.) is enough to provide component identification. In addition, those details were modeled in the operation concept and were not linked to the identification one. After discussion with them, they understood the identification risks and agreed about the limitations of only considering the technical details. Figure 8e shows also the highest recall (1) reached when dealing with the basic structure. This comes from the fact that our testers are experts in the power domain, hence they all chose the correct answers without forgetting any correct one. However, the lowest precision (of 0.575) was reached when dealing with the economic aspect, because some answered by choosing operational aspect parameters, since they considered that they are also related to the economic aspect.

- For electronic engineers: Figure 8f shows that the highest precision (of 1) obtained by our testers is also reached when dealing with the mobility aspect branch. However, the lowest precision (0.81) was reached when dealing with the operational aspect. This comes from the fact that the electricians are not all familiar with the operational and technical concepts of a power system. Figure 8f shows that the highest recall (of 1) is reached when dealing with the basic structure. This comes from the fact that most of them were not aware of all the details in the 'SG' domain. Hence, they chose almost all the proposed answers to avoid forgetting any correct one. However, the lowest precision (of 0.5) was reached when dealing with the operational aspect. This comes from the numerous correct answers, since testers focused on what they considered the most pertinent ones.

In order to consolidate the validation of our ontology structure, an additional experiment was added. In [1], the authors define consistency as a criterion that verifies if the ontology includes or allows any contradictions and propose the following SPARQL queries that search for anti-patterns, a strong indicator of in-consistencies, in the ontology. The first query detects concepts with no parent (cf. Fig. 9), and the second detects abnormally disjointed concepts in the ontology (cf. Fig. 10): We executed both queries and found no inconsistencies in our SSG ontology structure. This denotes the soundness of the integration of newly added concepts with the CIM and IEC standards.

```
SELECT ?a where {
              ?a rdfs:subClassOf owl:Nothing
}
```

Fig. 9. Anti-pattern of subsuming nothing

```
SELECT distinct ?A ?B1 ?B2 ?C1 where {
              ?B1 rdfs:subClassOf ?A .
              ?B2 rdfs:subClassOf ?A .
              ?C1 rdfs:subClassOf ?B1 .
              ?C1 owl:disjointWith ?B2 .
}
```

Fig. 10. Anti-pattern of skewed partitions

4.3.3   Discussion

Those results show that our ontology provides promising results in term of correctness and consistency, reflecting the comprehensibility and the clarity of our ontology concepts and relations for the experts and non-experts.

4.4   Domain Coverage Results

The domain coverage criterion comes down to evaluate the context layer of SSG. This latter targets the ontology capability of modeling the properties allowing the power system to meet the end-users needs by executing corresponding services. Hence, in order to evaluate it, SSG has been deployed into two main projects: HIT2GAP and ISare as detailed below. SSG has been serialized into RDF/OWL and posted online[2].

4.4.1   Integrating SSG in HIT2GAP

The $HIT2GAP$[3] is an European joint collaboration research project (EU/H2020 Grant Agreement No: 680708) for developing a next generation building control tool for optimizing energy usage. The main objective of this project is to propose a new paradigm of an energy management platform for smart buildings. The project consortium is composed of 22 partners from 10 European countries. The $HIT2GAP$ platform relies on an ontology allowing different partners to query data so to extract some information and events (through a set of services) from a smart building data. Figure 11 shows an extract of the ontological data model used for modeling and storing data within the platform. It shows its alignment with several main standards:

- IFC[4]: to represent the building related concepts,
- SSN[5]: to represent the data acquired from the sensors, and

---

[2] http://spider.sigappfr.org/research-projects/ontomg/.

[3] http://www.hit2gap.eu.

[4] http://www.buildingsmart-tech.org/specifications/ifc-overview.

[5] https://www.w3.org/2005/Incubator/ssn/ssnx/ssn.

- *SSG*: to represent all the power system equipment since a smart building can be considered as an *SG*.

Related concepts are prefixed with ifc:, ssn:, and *SSG*:. As one can see, *SSG* is integrated as a backbone of the information model of HIT2GAP platform. The following concepts have been aligned with *HIT2GAP* ontology:

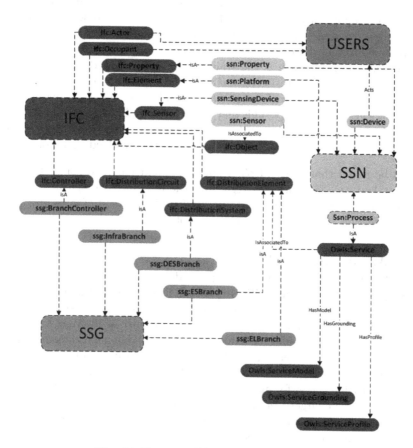

Fig. 11. Extract of HIT2GAP data model

1. *SSG*:DERBranch is aligned with ifc:DistributionSystem in order to extend the IFC with the distributed energy sources and their corresponding parameters,
2. *SSG*:ESBranch is aligned with ifc:DistributionElement in order to extend the IFC with the energy storage systems and their corresponding parameters,
3. *SSG*:ELBranch is aligned with ifc:DistributionElement in order to extend the IFC with the electrical loads and their corresponding parameters,
4. *SSG*:InfraBranch is aligned with ifc:DistributionCircuit in order to extend the IFC with the infrastructure equipment (e.g., cables, fiber optic, etc.) and their corresponding parameters,

5. SSG:`BranchController` is aligned with `ifc:Controller` in order to extend the IFC with the DES, ES, EL, Infrastructure controllers, and their corresponding parameters.

This alignment proves two main points:

- SSG is completely included in the HIT2GAP ontology since it allows to cover an important domain related to smart buildings: power domain. This will allow building actors to count on the expressiveness of SSG in order to represent/extract data and reason on it.
- SSG extends IFC which is the standard in building modelling that mainly focuses on the representation of the building equipment and constituents (e.g., floor, stair, wall, etc.), while neglecting the full coverage of the power related concepts in its vocabulary. This may weaken the building modeling since each equipment in the building can be considered as an energy source, storage or consumer, which highlights the importance of the SSG extension of the IFC.

It is to be noted that the $HIT2GAP$ project is currently on-going. Hence, we have not had any feedback yet regarding the domain coverage of SSG. The feedback of partners are expected to be received by the end of 2018 and will be posted online on the project website (See footnote 2).

4.4.2   Aligning SSG with ISare

In collaboration of Jema Irizar Group, leader of the ISare Microgrid (MG) project, we fully implemented SSG in it in order to highlight the potential of the ontology in answering the needs and objectives. ISare MG is installed in Spain and electrifies 12 offices. The generation system comprises 10 kW of solar generation, a nominal 53 kWh battery bank, 105 kW of wind generation and a 120 kW diesel genset. A second solar array of about 15 kW, mounted on the roof of the control system building, is connected to an SMA inverter and a 70 kWh of gas turbine to provide power for monitoring and communication. In addition, 50 kW of electric vehicle charger were installed, equipped with a protection system, to ensure a mobile power. The ISare MG has been modeled using our SSG, resulting the *ISare-SSGmodel*. As a power system, the ISare MG has several needs. ISare MG needs to be modeled via an interoperable structure, that enables the integration and the validation of the various new heterogeneous renewable distributed generation systems and various storage technologies. In order to enable ISare MG managers to have intuitive data querying and management, we developed a dedicated framework with an easy-to-use pool of predefined services so to achieve the objectives.

The *ISare-SSGmodel* has been implemented (cf. Fig. 12) as an OWL graph, on a central entity. Queries are executed through a SPARQL querying interface. Note that, SPARQL is a query language, that is, a semantic query language, able to retrieve and manipulate data stored in Web Ontology Language (OWL). Then, the HermiT reasoner has been added in order to interfere new knowledge and to allow the autonomous behavior of the MG. The idea behing choosing

HermiT is that it can determine whether or not the ontology is consistent and identify subsumption relationships between classes. In order to highlight the advantages provided by our *ISare-SSG*, three scenarios are presented in the following for illustration.

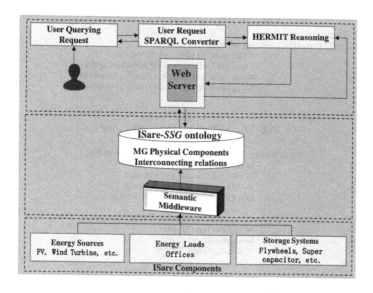

Fig. 12. ISare framework architecture

- Scenario 1 (Fig. 13): If an end-user needs to identify the consumer having the highest power consumption bill and advise him/her about the energy sources and storage systems that should be implemented in order to satisfy the demands at a lower cost, several concepts need to be used in the search engine from *ISare-SSG*. The basic-structure concepts are: *ELBranch*, *ESBranch* and *DERBranch*. Those of the Extended Structure are: Operation and Economic, with the following properties: *CptBill*, *EqCost*, *MaintenanceCost*, *InstallCost*, *OpCost*, *TotalCost*, *StrCost*, *StopCost*, *PwrKWhPrice*, *PwrhPrice* and *PwrCost*.
- Scenario 2 (Fig. 14): If an end-user needs to determine the most environmental friendly energy source, able to satisfy a consumer's power need at a certain weather condition, two basic-structure concepts are to be used: *ELBranch* and *DERBranch*, with other extended-structure concepts such as: Operation and ecology, with the following properties: *CarbEss*, *EthylEss*, *HeatEss*.
- Scenario 3 (Fig. 15): If an end-user wants to visualize the type, brand and model of the most implemented renewable energy sources (e.g., solar plant, wind plant, etc.) in the power system, his/her query will include the following basic-structure concepts: *ELBranch*, *DERBranch*, *ESBranch* and *InfraBranch*. It will also include one extended concept: Identification and all its properties (i.e., *Serial#*, *Type*, *Brand* and *Model*).

```
SELECT distinct ?A ?B WHERE {
        ?A rdfs:subClassOf ssg:ELBranch
        ?A ssg:hasCptBill max(ssg:CptBill)
        ?B rdfs:subClassOf ssg:ESBranch || ?B rdfs:subClassOf ssg:DERBranch
        ?B ssg:hasTotalCost min(ssg:TotalCost)
}
```

Fig. 13. Scenario 1 query example

```
SELECT distinct ?A WHERE {
        ?A rdfs:subClassOf ssg:DERBranch
        ?A ssg:hasCarbonEmission Min(ssg:CarbEss)
        ?A ssg:hasEthyleEmission Min(ssg:EthylEss)
        ?A ssg:hasEthyleEmission Min(ssg:HeatEss)
}
```

Fig. 14. Scenario 2 query example

```
SELECT distinct ?A ?Serial ?Type ?Brand ?Model WHERE {
        ?A rdfs:subClassOf ssg:DERBranch
        ?A rdfs:subClassOf ssg:ESBranch
        ?A rdfs:subClassOf ssg:ELBranch
        ?A rdfs:subClassOf ssg:InfraBranch

        ?A ssg:hasSerialNumber ?Serial
        ?A ssg:hasType ?Type
        ?A ssg:hasBrand ?Brand
        ?A ssg:hasModel ?Model
}
```

Fig. 15. Scenario 3 query example

4.4.3   Discussion

Those two applications show that our ontology provides a promising solid base for a better sharing of knowledge leading to a seamless communication between the components of the system (whether it is a smart building or a power system). In addition, it allows a better information querying and retrieval, and participates in increasing the reasoning capability of the system.

5   Conclusion

This paper introduces SSG, an ontology-based information model for SGs. The contributions of our work are four-folded: (1) it allows to resolve interoperability issues (syntactic and semantic) encountered between SG components, (2) it helps SG to represent and consider their (economical, ecological and operational) objectives directly in the information model (which is not the case of existing models) and allows to provide reasoning features to reach the fixed objectives, and (3) it allows to consider mobility and diversity of roles that can have each component involved in the SGs, and (4) it provides an evolutionary solution able to be extended easily to cover future needs. Several evaluations have been conducted to evaluate SSG resulting satisfactory results.

References

1. Brank, J., Grobelnik, M., Mladenic, D.: A survey of ontology evaluation techniques. In: Proceedings of the Conference on Data Mining and Data Warehouses (SiKDD 2005), pp. 166–170 (2005)
2. Catterson, V.M., Davidson, E.M., McArthur, S.D.: Issues in integrating existing multi-agent systems for power engineering applications. In: Proceedings of the 13th International Conference on, Intelligent Systems Application to Power Systems, 6-p. IEEE (2005)
3. Cox, W., Considine, T.: Energy, micromarkets, and microgrids. In: Grid-Interop 2011, pp. 1–8 (2011)
4. Cox, W., Holmberg, D., Sturek, D.: OASIS collaborative energy standards, facilities, and ZigBee smart energy. In: Grid-Interop Forum, pp. 1–8 (2011)
5. Cox, W.T., Considine, T., Principal, T.: Architecturally significant interfaces for the smart grid. In: Grid-Interop-The Road to an Interoperable Grid, Denver, Colorado, USA, pp. 17–19 (2009)
6. Dallinger, D., Wietschel, M.: Grid integration of intermittent renewable energy sources using price-responsive plug-in electric vehicles. Renew. Sustain. Energy Rev. **16**, 3370–3382 (2012)
7. BE Design: Facility smart grid information model (2017). https://www.iso.org/standard/71547.html. Accessed 30 July 2018
8. Fauzey, I.H.M., Nateghi, F., Mohammadi, F., Ismail, F.: Emergent occupational safety & health and environmental issues of demolition work: towards public environment. Procedia-Soc. Behav. Sci. **168**, 41–51 (2015)
9. Gillani, S., Laforest, F., Picard, G.: A generic ontology for prosumer-oriented smart grid. In: EDBT/ICDT Workshops, pp. 134–139 (2014)
10. Gómez-Pérez, A.: Ontology evaluation. In: Staab, S., Studer, R. (eds.) Handbook on Ontologies, pp. 251–273. Springer, Heidelberg (2004). https://doi.org/10.1007/978-3-540-24750-0_13
11. Goodwin, M., Yazidi, A.: A pattern recognition approach for peak prediction of electrical consumption. Integr. Comput.-Aided Eng. **23**, 101–113 (2016)
12. Grijalva, S., Costley, M., Ainsworth, N.: Prosumer-based control architecture for the future electricity grid. In: 2011 IEEE International Conference on Control Applications (CCA), pp. 43–48. IEEE (2011)
13. Guarino, N.: Formal ontology and information systems. In: Proceedings of FOIS, pp. 81–97. vol. 98 (1998)
14. Hammerstrom, D.J., et al.: Pacific northwest gridwise testbed demonstration projects. Part I. Olympic Peninsula Project, Technical report Pacific Northwest National Laboratory (PNNL), Richland, WA (US) (2007)
15. Hammerstrom, D.J., et al.: Pacific Northwest GridWise™ testbed demonstration projects; Part II. Grid Friendly™ Appliance Project. Technical report Pacific Northwest National Laboratory (PNNL), Richland, WA (US) (2007)
16. Ho, Q.-D., Le-Ngoc, T.: Smart grid communications networks: wireless technologies, protocols, issues and standards. In: Obaidat, M.S., Anpalagan, A., Woungang, I. (eds.) Handbook of Green Information and Communication Systems, pp. 115–146. Academic Press, New York (2012)
17. Kempton, W., Letendre, S.E.: Electric vehicles as a new power source for electric utilities. Transp. Res. Part D: Transp. Environ. **2**, 157–175 (1997)
18. Li, X., Liang, X., Lu, R., Shen, X., Lin, X., Zhu, H.: Securing smart grid: cyber attacks, countermeasures, and challenges. IEEE Commun. Mag. **50**(8), 38–45 (2012). https://doi.org/10.1109/MCOM.2012.6257525. ISSN 0163-6804

19. McMorran, A.W.: An introduction to IEC 61970–301 & 61968–11: The common information model. University of Strathclyde, 93, 124 (2007)
20. Mo, Y., et al.: Cyber-physical security of a smart grid infrastructure. Proc. IEEE **100**, 195–209 (2012)
21. Ritzer, G.: Focusing on the prosumer. In: Blättel-Mink, B., Hellmann, K.U. (eds.) Prosumer Revisited, pp. 61–79. Springer, Cham (2010)
22. Uschold, M., Gruninger, M.: Ontologies: principles, methods and applications. Knowl. Eng. Rev. **11**, 93–136 (1996)
23. Verhoosel, J., Rumph, F.-J., Konsman, M.: Modeling of flexibility in electricity demand and supply for renewables integration. In: Workshop on eeBuildings Data Models, Sophia Antipolis, France, pp. 1–8 (2011)
24. Wang, C., Liu, H., Wu, F.: The extend CIM for microgrid. In: 2012 China International Conference on Electricity Distribution (CICED), pp. 1–5 (2012). https://doi.org/10.1109/CICED.2012.6508543
25. Zowghi, D., Gervasi, V.: The three CS of requirements: consistency, completeness, and correctness. In: International Workshop on Requirements Engineering: Foundations for Software Quality, Essen, Germany: Essener Informatik Beitiage, pp. 155–164 (2002)

Bridging the Semantic Web and NoSQL Worlds: Generic SPARQL Query Translation and Application to MongoDB

Franck Michel$^{(\boxtimes)}$ ⓘ, Catherine Faron-Zucker ⓘ, and Johan Montagnat

Université Côte d'Azur, Inria, CNRS, I3S, Sophia Antipolis, France
`{franck.michel,johan.montagnat}@cnrs.fr, faron@i3s.unice.fr`

Abstract. RDF-based data integration is often hampered by the lack of methods to translate data locked in heterogeneous silos into RDF representations. In this paper, we tackle the challenge of bridging the gap between the Semantic Web and NoSQL worlds, by fostering the development of SPARQL interfaces to heterogeneous databases. To avoid defining yet another SPARQL translation method for each and every database, we propose a two-phase method. Firstly, a SPARQL query is translated into a pivot abstract query. This phase achieves as much of the translation process as possible regardless of the database. We show how optimizations at this abstract level can save subsequent work at the level of a target database query language. Secondly, the abstract query is translated into the query language of a target database, taking into account the specific database capabilities and constraints. We demonstrate the effectiveness of our method with the MongoDB NoSQL document store, such that arbitrary MongoDB documents can be aligned on existing domain ontologies and accessed with SPARQL. Finally, we draw on a real-world use case to report experimental results with respect to the effectiveness and performance of our approach.

Keywords: Query rewriting · SPARQL · RDF · NoSQL · xR2RML Linked data

1 Introduction

The Resource Description Framework (RDF) [11] is increasingly adopted as the pivot format for integrating heterogeneous data sources. It offers a unified data model that allows building upon countless existing vocabularies and domain ontologies, while benefiting from Semantic Web's reasoning capabilities. It also allows leveraging the growing, world-scale knowledge base referred to as the Web of Data. Today, increasing amounts of RDF data are published on the Web, notably following the Linked Data principles [2, 19]. These data often originate from heterogeneous silos that are inaccessible to data integration systems and search engines. Hence, a first step to enabling RDF-based data integration

A. Hameurlain et al. (Eds.): TLDKS XL, LNCS 11360, pp. 125–165, 2019.
https://doi.org/10.1007/978-3-662-58664-8_5

consists in translating legacy data from heterogeneous formats into RDF representations.

During the last fifteen years, much work has investigated how to translate common databases and data formats into RDF. Relational databases were primarily targeted [34,36], along with a handful of data formats such as XML [3] and CSV [28]. Meanwhile, the database landscape has significantly diversified with the adoption of various non-relational models. Initially designed as the core system of Big Data Web applications, NoSQL databases have gained momentum and are now increasingly adopted as general-purpose, commonplace databases. Today, companies and institutions store massive amounts of data in NoSQL instances. So far however, these data often remain inaccessible to RDF-based data integration systems, and consequently invisible to the Web of Data. although unleashing their data could potentially spur new integration opportunities and push the Web of Data forward.

The Semantic Web and NoSQL worlds build upon very different paradigms that are challenging to bridge over: whereas the former handles highly connected graphs along with the rich expressiveness of SPARQL, the latter trades off query expressiveness for scalability and fast retrieval of denormalized data[1]. As a result of these discrepancies, bridging the gap between those two worlds is a challenging endeavor.

Two strategies generally apply when it comes to access non-RDF data as RDF. In the *graph materialization* strategy, the transformation is applied exhaustively to the database content, the resulting RDF graph is loaded into a triple store and accessed through a SPARQL query engine [18] or by dereferencing URIs (as Linked Data). On the one hand, this strategy easily supports further processing or analysis, since the graph is made available at once. On the other hand, the materialized RDF graph may rapidly become outdated if the pace of database updates is high. Running the transformation process periodically is a common workaround, but in the context of large data sets, the cost (in time, memory and CPU) of materializing and reloading the graph may become out of reach. To work out this issue, the *query rewriting* strategy aims to access heterogeneous databases as virtual RDF graphs. A query processor rewrites a SPARQL query into the query language of the target database. The target database query is evaluated at run-time such that only relevant data are fetched from the database and translated into RDF triples. This strategy better scales to big data sets and guarantees data freshness, but entails overheads that may penalize performances if complex analysis is needed.

In previous works we defined a generic mapping language, xR2RML [25], that enables the translation of a broad scope of data sources into RDF. The mapping instructs how to translate each data item from its original format into RDF triples, by adapting to the multiplicity of query languages and data models. We

---

[1] We refer to key-value stores, document stores and column family stores but leave out graph stores that generally come with a richer query expressiveness.

applied xR2RML to the MongoDB NoSQL document store[2] and we implemented the *graph materialization* strategy.

To cope with large and frequently updated data sets though, we wish to tackle the question of accessing such databases using the query rewriting strategy. Hence, to avoid defining yet another SPARQL translation method for each and every database, in this paper we investigate a general two-phase method. Firstly, given a set of xR2RML mappings, a SPARQL query is rewritten into a pivot abstract query. This phase achieves as much of the translation process as possible regardless of the database, and enforces early query optimizations. Secondly, the abstract query is translated into the target database query language, taking into account the specific database capabilities and constraints. We demonstrate the effectiveness of our method in the case of MongoDB, accessing arbitrary MongoDB documents with SPARQL. We show that we can always rewrite an abstract query into a union of MongoDB *find* queries that shall return all the documents required to answer the SPARQL query.

The rest of this article is organized as follows. After a review of SPARQL query rewriting approaches in Sect. 2, we quickly remind the principles and main features of the xR2RML mapping language in Sect. 3. Then, in Sects. 4 and 5 we describe the two-phase method introduced above. In Sect. 6, we describe a real-world use case and we report experimental results with respect to the effectiveness and performance of our approach. Finally, we discuss our solution and envision some perspectives in Sect. 7, and we draw some conclusions in Sect. 8.

2   Related Works

2.1   Rewriting SPARQL to SQL and XQuery

Since the early 2000's, various works have investigated methods to query legacy data sources with SPARQL. Relational databases (RDB) have caught much attention, either in the context of RDB-backed RDF stores [10,14,35] or using arbitrary relational schemas [5,29,31,32,38]. These methods harness the ability of SQL to support joins, unions, nested queries and various string manipulation functions. Typically, a conjunction of two SPARQL basic graph patterns (BGP) results in the inner join of their respective translations; their union results in a SQL UNION ALL clause; the SPARQL OPTIONAL clause results in a left outer join, and a SPARQL FILTER results in an encapsulating SQL SELECT WHERE clause.

Chebotko's algorithm [10] focused on RDB-based triple stores. Priyatna et al. [29] extended it to support custom R2RML mappings (the W3C recommendation of an RDB-to-RDF mapping language [12]) while applying several query optimizations. Two limitations can be emphasized though: (i) R2RML mappings must have constant predicates, *i.e.* the predicate term of the generated

---

[2] https://www.mongodb.org/.

RDF triples cannot be built from database values; (ii) Triple patterns are considered and translated independently of each other, even when they share SPARQL variables. The resulting SQL query embeds unnecessary complexity that is taken care of later on, in the SQL query optimization step. Unbehauen et al. [38] clear the first limitation by defining the concept of compatibility between the RDF terms of a SPARQL triple pattern and R2RML mappings, which enables managing variable predicates. Furthermore, to address the second limitation, they pre-checking join constraints implied by shared variables in order to reduce the number of candidate mappings for each triple pattern. Yet again, two limitations can be noticed: (iii) References between R2RML mappings are not considered, hence joins implied by shared variables are dealt with but joins declared in the R2RML mapping graph are ignored. (iv) The rewriting process associates each part of a mapping to a set of columns, called column group, which enables filter, join and data type compatibility checks. This leverages SQL capabilities (CASE, CAST, string concatenation, etc.), making it hardly applicable out of the scope of SQL-based systems. In the three aforementioned approaches, the optimization is dependent on the target database language, and can hardly be generalized. In our attempt to rewrite SPARQL queries in the general case, such optimization are performed earlier, regardless of the target database capabilities.

In a somewhat different approach, Rodríguez-Muro and Rezk [32] extend the *ontop* Ontology-Based Data Access (OBDA) system to support R2RML mappings. A SPARQL query and an R2RML mapping graph are translated into a Datalog program. This formal representation is used to combine and apply optimization techniques from logic programming and SQL querying. The optimized program is then translated into an executable SQL query.

Other approaches investigated the querying of XML databases in a rather similar philosophy. For instance, SPARQL2XQuery [4] relies on the ability of XQuery to support joins, nested queries and complex filtering. Typically, a SPARQL FILTER is translated into an encapsulating For-Let-Where XQuery clause.

Finally, it occurs that the rich expressiveness of SQL and XQuery makes it possible to translate a SPARQL 1.0 query into a single, possibly deeply nested, target query, whose semantics is provably strictly equivalent to that of the SPARQL query. Commonly, query optimization issues are addressed at the level of the produced target query, or they may even be delegated to the target database optimization engine. Hence, the above reviewed methods are tailored to the expressiveness of the target query language, such that SQL or XQuery specificities are woven into the translation method itself, which undermines the ability to use such methods beyond their initial scope.

2.2   Rewriting SPARQL to NoSQL

To the best of our knowledge, little work has investigated how to perform RDF-based data integration over the NoSQL family of databases. An early work[3]

---

[3] https://github.com/agrueneberg/Sessel.

has tackled the translation of CouchDB[4] documents into RDF, but did not addressed SPARQL rewriting. MongoGraph[5] is an extension of the AllegroGraph triple store to query arbitrary MongoDB documents with SPARQL. But very much like the Direct Mapping [1] defined in the context of RDBs, both works come up with an ad-hoc ontology (*e.g.* each JSON field name is turned into a predicate) and hardly supports the reuse of existing ontologies. Tomaszuk proposed to use a MongoDB database as an RDF triple store [37]. In this context, the author devised a translation of SPARQL queries into MongoDB queries, that is however closely tied to the specific database schema and thus is unfit for arbitrary documents.

More in line with our work, Botoeva et al. proposed a generalization of the OBDA principles [30] to MongoDB [8]. They describe a two-step rewriting process of SPARQL queries into a MongoDB *aggregate* pipeline. In Sect. 7, we analyze in further details the relationship between their approach and ours. Interestingly, to the best of our knowledge, only one approach tackled the key-value store subset of NoSQL databases. Mugnier et al. [26] define the NO-RL rule language that can express lightweight ontologies to be applied to key-value stores. Leveraging the formal semantics of NO-RL, they propose an algorithm to reformulate a query under a NO-RL ontology, but SPARQL is not considered.

Finally, since NoSQL document stores are based on JSON, let us mention the JSON-LD syntax that is meant for the serialization of Linked Data in the JSON format. When applied to existing JSON documents, a JSON-LD profile can be considered as a lightweight method to interpret JSON data as RDF. Such a profile could be exploited by a SPARQL rewriting engine to enable the querying of document stores with SPARQL. This approach would be limited though, since JSON-LD is not meant to describe rich mappings from JSON to RDF, but simply to interpret JSON as RDF. It lacks the expressiveness and flexibility required to align JSON documents with domain ontologies that may model data in a rather different manner. Besides, we do not want to define a method specifically tailored to MongoDB; our point is to provide a generic rewriting method that can be applied to the concrete case of MongoDB as well as various other databases.

3 The xR2RML Mapping Language

The xR2RML mapping language [25] intends to foster the translation of legacy data sources into RDF. It can describe the mapping of an extensible scope of databases to RDF, independently of any query language or data model. It is backward compatible with R2RML and relies on RML [13] for the handling of various data formats. It can translate data with mixed embedded formats and generate RDF collections and containers.

An xR2RML mapping defines a logical source (property `xrr:logicalSource`) as the result of executing a query against an input database (`xrr:query` and

---

[4] http://couchdb.apache.org/.

[5] http://franz.com/agraph/support/documentation/4.7/mongo-interface.html.

`rr:tableName`). An optional iterator (value of property `rml:iterator`) can be applied to each query result, and a `xrr:uniqueRef` property can identify unique fields. Data from the logical source is mapped to RDF terms (literal, IRI, blank node) by term maps. There exists four types of term maps: a subject map generates the subject of RDF triples, predicate and object maps produce the predicate and object terms, and an optional graph map is used to name a target graph. Listing 1.1 depicts two mappings `<#Mbox>` and `<#Knows>`, each consisting of a subject map, a predicate map and an object map.

Term maps extract data from query results by evaluating *xR2RML references* whose syntax depends on the target database and is an implementation choice: typically, this may be a column name in case of a relational database, an XPath expression in case of an XML database, or a JSONPath[6] expression in case of NoSQL document stores like MongoDB or CouchDB. xR2RML references are used with property `xrr:reference` whose value is a single xR2RML reference, and property `rr:template` whose value is a template string which may contain several references. In Listing 1.1, both subject maps use a template to build IRI terms by concatenating http://example.org/member/ with the value of the `"id"` JSON field.

```
<#Mbox>
  xrr:logicalSource [ xrr:query "db.people.find({'emails':{$ne: null}})" ];
  rr:subjectMap [ rr:template "http://example.org/member/{$.id}" ];
  rr:predicateObjectMap [
    rr:predicate foaf:mbox;
    rr:objectMap [ rr:template "mailto:{$.emails.*}"; rr:termType rr:IRI ]
  ].
<#Knows>
  xrr:logicalSource [
    xrr:query "db.people.find({'contacts':{$size: {$gte:1}}})" ];
  rr:subjectMap [ rr:template "http://example.org/member/{$.id}" ];
  rr:predicateObjectMap [
    rr:predicate foaf:knows;
    rr:objectMap [
      rr:parentTriplesMap <#Mbox>;
      rr:joinCondition [ rr:child "$.contacts.*"; rr:parent "$.emails.*" ] ]
  ].
```

Listing 1.1. xR2RML example mapping graph

When the evaluation of an xR2RML reference produces several RDF terms, the xR2RML processor creates one triple for each term. Alternatively, the `rr:termType` property of a term map can be used to group the terms in an RDF collection while specifying a language tag or data type. Besides, the default iteration model can be modified using *nested term maps*, notably useful to parse nested collections of values and generate appropriate triples.

xR2RML allows to model cross-references by means of *referencing object maps* that use values produced by the subject map of a parent mapping as the objects of triples produced by a child mapping. Properties `rr:child` and `rr:parent` specify the join condition between documents of both mappings.

Running Example. To illustrate the description of our method, we define a running example that we shall use throughout this paper. Let us consider a

---

[6] http://goessner.net/articles/JsonPath/.

```
{ "id":            105632,
  "firstname":"John",
  "emails":     ["john@foo.com","john@example.org"],
  "contacts":   ["chris@example.org", "alice@foo.com"] }

{ "id":            327563,
  "firstname":"Alice",
  "emails":     ["alice@foo.com"],
  "contacts":   ["john@foo.com"] }
```

Listing 1.2. MongoDB collection "people" containing two documents

MongoDB database with a collection **people** depicted in Listing 1.2: each JSON document provides the identifier, email addresses and contacts of a person; contacts are identified by their email addresses.

Let us now consider the xR2RML mapping graph in Listing 1.1, consisting of two mappings <#Mbox> and <#Knows>. The logical source of mappings <#Mbox>, respectively <#Knows>, is a MongoDB query that retrieves documents having a non-null **emails** field, respectively a **contacts** array field with at least one element. Both subject maps use a template to build IRI terms by concatenating http://example.org/member/ with the value of JSON field **id**. Applied to the documents in Listing 1.2, the xR2RML mapping graph generates the following RDF triples:

```
<http://example.org/member/105632>
  foaf:mbox <mailto:john@foo.com>, <mailto:john@example.org>;
  foaf:knows <http://example.org/member/327563>.

<http://example.org/member/327563>
  foaf:mbox <mailto:alice@foo.com>;
  foaf:knows <http://example.org/member/105632>.
```

4   From SPARQL to Abstract Queries

Section 2 emphasized that SPARQL rewriting methods for SQL or XQuery rely on prior knowledge about the target query language expressiveness. This makes possible the semantics-preserving translation of a SPARQL query into a single equivalent target query. In the general case however (beyond SQL and XQuery), the target query language may not support joins, unions, sub-queries and/or filtering. To tackle this challenge, our method first enacts the database-independent steps of the rewriting process. To generate the abstract query, we rely on and extend the R2RML-based SPARQL rewriting approaches reviewed in Sect. 2, while taking care of avoiding the limitations highlighted. More specifically, we focus on rewriting a SPARQL 1.0 graph pattern, whatever the query form (SELECT, ASK, DESCRIBE, etc.). The translation of a SPARQL graph pattern into an abstract query consists of four steps, sketched in Fig. 1 and described in the next sub-sections. Sect. 4.1: A SPARQL 1.0 graph pattern is rewritten into an abstract expression exhibiting operators of the abstract query language. Sect. 4.2: We identify candidate xR2RML mappings likely to generate RDF triples that match each triple pattern. Sect. 4.3: Each triple pattern is translated into a sub-query according to the set of xR2RML mappings identified. A sub-query consists of operators of the abstract query language and atomic abstract queries. Sect. 4.4: We enforce several optimizations on the resulting abstract query, *e.g.* self-joins or self-unions elimination.

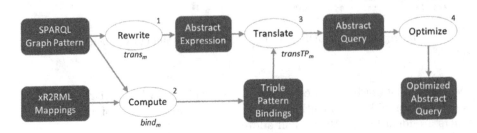

Fig. 1. Translation of a SPARQL 1.0 graph pattern into an optimized abstract query

4.1   Translation of a SPARQL Graph Pattern

Our pivot abstract query language complies with the grammar depicted in Definition 1. It derives from the syntax and semantics of SPARQL [27]: the language keeps the names of several SPARQL operators (UNION, LIMIT, FILTER) and prefers the SQL terms INNER JOIN ON and LEFT OUTER JOIN ON to refer to join operations more explicitly. A notable difference with SPARQL is that, in the tree representation of a query, the leaves of a SPARQL query are triple patterns. Conversely, the leaves of an abstract query are *Atomic Abstract Queries* (Sect. 4.3).

The INNER JOIN and LEFT OUTER JOIN operators stem from the join constraints implied by shared variables. Somehow, the second INNER JOIN in Definition 1, including the *"AS child"* and *"AS parent"* notations, is entailed by the join constraints expressed in xR2RML mappings using referencing object maps and properties `rr:child` and `rr:parent`. Notation $v_1, ...v_n$, in the join operators, stands for the set of SPARQL variables on which the join is to be performed. Notation $<Ref>$ stands for any valid xR2RML data element reference, *i.e.* a column name for a tabular data source, an XPath expression for an XML database, a JSONPath expression for a NoSQL document store such as MongoDB and CouchDB, etc.

Definition 1. *Grammar of the Abstract Pivot Query Language*

```
<AbstractQuery> ::= <AtomicQuery> | <Query> |
                <Query> FILTER <SPARQL filter> | <Query> LIMIT <integer>
<Query> ::= <AbstractQuery> INNER JOIN <AbstractQuery> ON {v₁,...vₙ} |
            <AbstractQuery> AS child INNER JOIN <AbstractQuery> AS parent
                ON child/<Ref> = parent/<Ref> |
            <AbstractQuery> LEFT OUTER JOIN <AbstractQuery> ON {v₁,... vₙ}|
            <AbstractQuery> UNION <AbstractQuery>
<AtomicQuery> ::= {From, Project, Where, Limit}
```

The first query transformation step is implemented by function $trans_m$ depicted in Definition 2. It rewrites a well-designed SPARQL graph pattern [27] into an abstract query while making no assumption with respect to the target database query capabilities. It extends the algorithms proposed in [10,29,38].

Definition 2. Translation of a SPARQL query into an abstract query under xR2RML mappings (functiontrans$_m$).
Let m be an xR2RML mapping graph consisting of a set of xR2RML mappings. Let gp be a well-designed SPARQL graph pattern, f be a SPARQL filter and l an integer limit value representing the maximum number of results.
We denote by $trans_m(gp, f, l)$ *the translation, under m, of "gp FILTER f" into an abstract query that shall not return more than l results. We denote by* $trans_m(gp)$ *the result of* $trans_m(gp, true, \infty)$. *Function* $trans_m$ *is defined recursively as follows:*

- *if gp consists of a single triple pattern tp,* $trans_m(gp, f, l) = transTP_m(tp,$ *sparqlCond(tp, f), l)*
  where $transTP_m$ *translates a single triple pattern into an abstract query (Sect. 4.3) and* **sparqlCond** *discriminates SPARQL filter conditions (Sect. 4.1).*
- *if gp is (P LIMIT l'),* $trans_m(gp, f, l) = trans_m(gp, f, min(l, l'))$
- *if gp is (P FILTER f'),* $trans_m(gp, f, l) = trans_m(P, f \wedge f', \infty)$ *FILTER sparqlCond(P, f ∧ f') LIMIT l*
- *if gp is* $(P_1$ *AND* $P_2)$, $trans_m(gp, f, l) = trans_m(P_1, f, \infty)$ *INNER JOIN* $trans_m(P_2, f, \infty)$ *ON* $var(P_1) \cap var(P_2)$ *LIMIT l*
- *if gp is* $(P_1$ *OPTIONAL* $P_2)$, $trans_m$ $(gp, f, l) =$
  $trans_m(P_1, f, \infty)$ *LEFT OUTER JOIN* $trans_m(P_2, f, \infty)$ *ON* $var(P_1) \cap$ $var(P_2)$ *LIMIT l*
- *if gp is* $(P_1$ *UNION* $P_2)$, $trans_m(gp, f, l) = trans_m(P_1, f, l)$ *UNION* $trans_m(P_2, f, l)$ *LIMIT l*

As a simplification, notations "FILTER true" and "LIMIT ∞" may be omitted.

Example. Let us give a first simple illustration. SPARQL query Q_1 contains a graph pattern gp_1 that consists of two triple patterns, tp_1 and tp_2:

```
Q₁: SELECT ?x WHERE {
       ?x foaf:mbox ?mbox. # tp₁
       ?x foaf:knows ?y. } # tp₂
```

The application of function $trans_m$ to the graph pattern gp_1 is as follows:

```
transₘ(gp₁)
= transₘ(gp₁, true, ∞)
= transTPₘ(tp₁, true, ∞) INNER JOIN
  transTPₘ(tp₂, true, ∞) ON {var(tp₁) ∩ var(tp₂)}
  LIMIT ∞
= transTPₘ(tp₁) INNER JOIN transTPₘ(tp₂) ON {?x}
```

Dealing with SPARQL Filters. SPARQL rewriting methods reviewed in Sect. 2 generally adopt a bottom-up approach where, typically, a SPARQL FILTER translates into an encapsulating query (*e.g.* a SELECT-WHERE clause in the case of SQL). Thus, filters in the outer query do not contribute to the selectivity of inner-queries that may return large intermediate results. This flaw is commonly worked out in a subsequent SQL query optimization step, or by assuming that the underlying database engine can take care of this optimization.

In our context though, we cannot assume that the target query can be optimized nor that the database query engine is capable of doing it. We therefore consider SPARQL filters at the earliest stage: function $trans_m$ pushes SPARQL filters down into the translation of each inner query in order to return only necessary intermediate results.

Let us consider a SPARQL filter f as a conjunction of n conditions ($n \geq 1$): $C_1 \wedge ... C_n$. Function *sparqlCond*, formally defined in [22], discriminates between these conditions with regards to two criteria:

(i) A condition C_i is pushed into the translation of triple pattern tp if all variables of C_i show up in tp, *e.g.* a condition involving variables $?x$ and $?y$ is pushed into the translation of tp only if tp involves at least $?x$ and $?y$.

(ii) A condition C_i is part of the abstract FILTER operator if at least one variable of C_i is shared by several triple patterns, *e.g.* if C_i contains variable $?x$, and variable $?x$ also shows up in two different triple patterns, then C_i is in the condition of the abstract FILTER operator.

Note that both criteria are not exclusive: a condition may simultaneously show up in the translation of a triple pattern and in the FILTER abstract operator.

Example. SPARQL query Q_2, depicted in Listing 1.3, contains the graph pattern gp_2 that consists of three triple patterns tp_1, tp_2 and tp_3, and a filter consisting of the conjunction of two conditions c_1 and c_2:

```
SELECT ?x WHERE {
    ?x foaf:mbox ?mbox.                      # tp1
    ?y foaf:mbox <mailto:john@foo.com>.      # tp2
    ?x foaf:knows ?y.                        # tp3
    FILTER {
        contains(str(?mbox), "foo.com")      # c1
        && ?x != ?y } }                      # c2
```

Listing 1.3. SPARQL query Q_2

Let us compute function *sparqlCond* for each triple pattern:

- tp_1 has two variables, $?x$ and $?mbox$. No condition involves both variables, but c_1 involves $?mbox$ and has no other variable, thereby c_1 matches criterion (i) for tp_1. Condition c_2 involves $?x$ but it also involves $?y$ that is not in tp_1. Hence, c_2 does not match criterion (i) for tp_1, and sparqlCond(tp_1, $c_1 \wedge c_2$) = c_1.

- tp_2 has one variable, $?y$, and no condition involves only $?y$. Hence, no condition can be pushed into the translation of tp_2, denoted sparqlCond(tp_2, $c_1 \wedge c_2$) = true.

- tp_3 has two variables $?x$ and $?y$, and only condition c_2 involves them both. Hence, only c_2 matches criterion (i) for tp_3 and $\mathtt{sparqlCond(tp_3,\ c_1 \wedge c_2)\ =\ c_2}$.
- Lastly, only condition c_2 involves variables shared by several triples patterns: $?x$ and $?y$. Thus, only c_2 matches criterion (ii), which entails the generation of the abstract filter $\mathtt{FILTER(c_2)}$.

As a result, gp_2 is rewritten into the following abstract query:

```
transm(gp2, c1 ∧ c2) = transTPm(tp1, c1)
  INNER JOIN transTPm(tp2, true) ON {}
  INNER JOIN transTPm(tp3, c2) ON {?x,?y}
  FILTER(c2)
```

Dealing with the LIMIT Solution Modifier. Similar to the case of SPARQL filters, the common bottom-up approach of SQL rewriting methods consists in rewriting a LIMIT into an encapsulating query. Thus, again, sub-queries may return unnecessary large intermediate results. Therefore, function $trans_m$ pushes the LIMIT value down into the translation of each triple pattern using the limit argument l, initialized to ∞. During the parsing of the graph pattern by function $trans_m$, the limit argument is updated according to the graph pattern encountered. Below, we elaborate on some of the situations tackled in Definition 2:

- In a graph pattern $P\ LIMIT\ l'$, the smallest limit is kept, hence the $min(l, l')$ in $\boldsymbol{trans_m}(gp,\ f,\ min(l,\ l'))$.
- In a graph pattern $P\ FILTER\ f'$, we cannot know in advance how many results will be filtered out by the FILTER clause. Consequently, we have to run the query with no limit and apply the filter afterward. Hence the ∞ argument in $\boldsymbol{trans_m}(P,\ f \wedge f',\ \infty)\ \boldsymbol{FILTER\ sparqlCond(...)}\ \boldsymbol{LIMIT}\ l$.
- Similarly, in the case of an inner or left join, we cannot know in advance how many results will be returned. Consequently, the left and right queries alike are run with no limit first, the join is computed, and only then can we limit the number of results. Hence the ∞ argument in the expressions: $\boldsymbol{trans_m}(P_1, f, \infty)\ ...\ \boldsymbol{INNER\ JOIN}\ \boldsymbol{trans_m}(P_2, f, \infty)\ ...\ \boldsymbol{LIMIT}\ l$.

Dealing with Other Solution Modifiers. For the sake of simplicity, we do not describe in further details the management of SPARQL solution modifiers OFFSET, ORDER BY and DISTINCT. Let us simply mention that they are managed in the very same way as the SPARQL FILTER clause and LIMIT solution modifier, *i.e.* as additional parameters of the $trans_m$ and $transTP_m$ functions, and additional operators of the abstract query language.

4.2   Binding xR2RML Mappings to Triple Patterns

An important step in the rewriting process consists in figuring out which of the mappings are good candidates to answer the SPARQL query. More precisely, for each triple pattern tp of the SPARQL graph pattern, we must figure out which mappings can possibly generate triples that match *tp*. We call this the *triple pattern binding*[7], defined in Definition 3:

Definition 3. *Triple Pattern Binding.*
*Let m be an xR2RML mapping graph consisting of a set of xR2RML mappings, and tp be a triple pattern. A mapping $M \in m$ is **bound** to tp if it is likely to produce triples that match tp. A **triple pattern binding** is a pair (tp, MSet) where MSet is the set of mappings of m that are bound to tp.*

Function *bind$_m$* (Definition 4) determines, for a graph pattern *gp*, the bindings of each triple pattern of *gp*. It takes into account join constraints implied by shared variables and by cross-references defined in the mapping (xR2RML referencing-object map), and the SPARQL filter constraints whose unsatisfiability can be verified statically. This is achieved by means of two functions: *compatible* and *reduce*. These functions were introduced in [38] but important details were left untold. Especially, the authors did not formally define what the compatibility between a term map and a triple pattern term means, and they did not investigate the compatibility between a term map and a SPARQL filter. In this section we give a detailed insight into these functions. A formal definition is provided in [22].

Definition 4. *Binding xR2RML mappings to triple patterns (bind$_m$).*
Let m be a set of xR2RML mappings, gp be a well-designed graph pattern, and f be a SPARQL filter. Let M.sub, M.pred and M.obj respectively denote the subject map, the predicate map and the object map of an xR2RML mapping M.
*We denote by **bind$_m$**(gp, f) the set of triple pattern bindings of "gp FILTER f" under m, and we denote by **bind$_m$**(gp) the result of **bind$_m$**(gp, true).*
*Function **bind$_m$**(gp, f) is defined recursively as follows:*

- *if gp consists of a single triple pattern tp, **bind$_m$**(gp, f) is the pair (tp, MSet) where MSet = {M | M ∈ m ∧ **compatible**(M.sub, tp.sub, f) ∧ **compatible**(M.pred, tp.pred, f) ∧ **compatible**(M.obj, tp.obj, f)}*
  *where **compatible** verifies the compatibility between a term map, a triple pattern term and a SPARQL filter*
- *if gp is (P1 AND P2), **bind$_m$**(gp, f) = **reduce**(**bind$_m$**(P1, f), **bind$_m$**(P2, f)) ∪ **reduce**(**bind$_m$**(P2, f), **bind$_m$**(P1, f))*
  *where **reduce** utilizes dependencies between graph patterns to reduce their bindings*
- *if gp is (P1 OPTIONAL P2), **bind$_m$**(gp, f) = **bind$_m$**(P1, f) ∪ **reduce**(**bind$_m$**(P2, f), **bind$_m$**(P1, f))*
- *if gp is (P1 UNION P2), **bind$_m$**(gp, f) = **bind$_m$**(P1, f) ∪ **bind$_m$**(P2, f)*
- *if gp is (P FILTER f'), **bind$_m$**(gp, f) = **bind$_m$**(P, f ∧ f')*

---

[7] We adapt the *triple pattern binding* proposed by Unbehauen et al. in [38], and we assume that xR2RML mappings are *normalized* in the sense defined by [32], i.e. they contain exactly one predicate-object map with exactly one predicate map and one object map, and any `rr:class` property is replaced by an equivalent predicate-object map with a constant predicate `rdf:type`.

Function *compatible* checks whether a term map is compatible with (i) a term of a triple pattern and (ii) a SPARQL filter, so as to rule out incompatible associations. When the triple pattern term is constant (literal, IRI or blank node), incompatibilities may occur when its type does not mach the term map type (*e.g.* when the triple pattern term is a literal whereas the term map produces IRIs). Incompatibilities may also occur for literals when language tags or data types do not match. When the triple pattern term is a variable, incompatibilities may arise from unsatisfiable SPARQL filters. These situations pertain to type constraints expressed using SPARQL functions *isIRI*, *isLiteral* or *isBlank*, as well as language and data type constraints expressed using functions *lang*, *langMatches* and *datatype*. For instance, if variable $?v$ is associated with a term map that produces literals, the SPARQL filter *isIRI($?v$)* can never be satisfied, which ensures that the association is invalid. We provided a formal definition of function *compatible* in [23].

Function *reduce* uses the variables shared by two triple patterns to detect unsatisfiable join constraints, and accordingly to reduce the set of mappings bound to each triple pattern. For instance, let us consider two triple patterns tp_1 and tp_2 that have a shared variable $?v$. Mapping M_1 is bound to tp_1 and mapping M_2 is bound to tp_2. If the term map associated to $?v$ in M_1 generates literals whereas the term map associated to $?v$ in M_2 generates IRIs, we say that the term maps are incompatible. Consequently, function *reduce* rules out M_1 from the bindings of tp_1 and M_2 from the bindings of tp_2. In other words, *reduce($bind_m(tp_1)$, $bind_m(tp_2)$)* returns the reduced bindings of tp_1 such that the term maps associated to $?v$ in the bindings of tp_1 are compatible with the term maps associated to $?v$ in the bindings of tp_2.

Running Example. Let us consider query Q_2 depicted in Listing 1.3. We first compute the triple pattern bindings for tp_1, tp_2 and tp_3 independently. The constant predicate of tp_1 and tp_2 matches the constant predicate map of mapping `<#Mbox>`. The subject and object of tp_1 are both variables, and the constant object of tp_2 (<mailto:john@foo.com>) is compatible with the object map of `<#Mbox>`. Hence, `<#Mbox>` is bound to both triple patterns:

$bind_m(tp_1, c_1 \wedge c_2) = (tp_1, \{$`<#Mbox>`$\})$
$bind_m(tp_2, c_1 \wedge c_2) = (tp_2, \{$`<#Mbox>`$\})$

Likewise, we can show that `<#Knows>` is bound to tp_3:

$bind_m(tp_3, c_1 \wedge c_2) = (tp_3, \{$`<#Knows>`$\})$.

Let us consider the join constraint implied by variable $?y$:

```
?y foaf:mbox <mailto:john@foo.com>.   # tp2
?x foaf:knows ?y.                     # tp3
```

$?y$ is the subject in tp_2 that is bound to `<#Mbox>`, $?y$ is thereby associated to `<#Mbox>`'s subject map. $?y$ is also the object in tp_3 that is bound to `<#Knows>`, $?y$ is thereby associated to `<#Knows>`'s object map. Therefore, the expression

$reduce(bind_m(tp_2, c_1 \wedge c_2), bind_m(tp_3, c_1 \wedge c_2))$

checks whether the subject map of `<#Mbox>` is compatible with the object map of `<#Knows>`. But since the object map of `<#Knows>` is a referencing object map whose parent is `<#Mbox>`, this amounts to check whether the subject map of `<#Mbox>` is compatible with itself, which is obvious. Consequently, the join constraint implied by variable $?y$ does not rule out any binding.

Similarly, we can show that the join constraint implied by variable $?x$, shared by tp_1 and tp_3, does not rule out any binding. Lastly, the set of triple pattern bindings for the graph pattern of query Q_2 is as follows:

$bind_m$(tp$_1$ AND tp$_2$ AND tp$_3$, $c_1 \wedge c_2$) =
(tp$_1$,{`<#Mbox>`}), (tp$_2$,{`<#Mbox>`}), (tp$_3$,{`<#Knows>`})

4.3 Translation of a SPARQL Triple Pattern

The last step of the rewriting towards the abstract query language consists in the translation of each triple pattern into an abstract query, under the set of xR2RML mappings bound to that triple pattern by function $bind_m$. This is achieved by function $transTP_m$ defined in Definition 5, that may have to deal with various situations.

Definition 5. *Translation of a SPARQL Triple Pattern into Atomic Abstract Queries (function transTP$_m$).*
Let m be an xR2RML mapping graph consisting of a set of xR2RML mappings, gp be a well-designed graph pattern, and tp a triple pattern of gp. Let l be the maximum number of query results, and f be a SPARQL filter expression. Let getBoundM$_m$(gp, tp, f) be the function that, given gp, tp and f, returns the set of mappings of m that are bound to tp in bind$_m$(gp, f).
We denote by **transTP$_m$***(tp, f, l) the translation, under getBoundM$_m$(gp, tp, f), of tp into an abstract query whose results can be translated into at most l RDF triples matching "tp FILTER f". The resulting abstract query, denoted* `<ResultQuery>` *in the grammar below, is a union of per-mapping subqueries, where a subquery is either an Atomic Abstract Query or the inner join of two Atomic Abstract Queries.*
As a simplification, arguments f and l may be omitted when their values are "true" and ∞ respectively.

```
<ResultQuery> ::= <SubQuery> (UNION <SubQuery>)*
<SubQuery>    ::= <AtomicQuery> |
                  <AtomicQuery> AS child INNER JOIN <AtomicQuery> AS parent
                  ON child/<Ref>=parent/<Ref>
```

Let us now give an insight into how $transTP_m$ deals with these situations.

(1) The most simple situation is encountered when a simple triple pattern tp is bound with a single xR2RML mapping M. If M has a regular object map (not a referencing object map denoting a cross-reference), then tp translates into an *atomic abstract query*. We will define the concept of atomic abstract query further on in this section. At this point, let us just notice that it is an abstract query obtained by matching the terms of a triple pattern with their respective term maps in a mapping.

(2) If the mapping M denotes a cross-reference by means of a referencing object map, *i.e.* it refers to another mapping for the generation of object terms, then the result of $transTP_m$ is the INNER JOIN of two atomic abstract queries, denoted:

```
<AtomicQuery1> AS child INNER JOIN
<AtomicQuery2> AS parent ON
child/childRef=parent/parentRef
```

where **childRef** and **parentRef** denote the values of properties `rr:child` and `rr:parent` respectively.

(3) We have seen, in the definition of $bind_m$, that several mappings may be bound to a single triple pattern tp, each one may produce a subset of the RDF triples that match tp. In such a situation, $transTP_m$ translates tp into a UNION of per-mapping atomic abstract queries.

Interestingly enough, we notice that INNER JOINs may be implied either by shared SPARQL variables (Definition 2) or cross-references denoted in the mappings (situation (2) described above). Similarly, UNIONs may arise either from the SPARQL UNION operator (Definition 2) or the binding of several mappings to the same triple pattern (situation (3) described above).

Due to size constraints, we do not go through the full algorithm of $transTP_m$ in this paper, however the interested reader is referred to [22] for a comprehensive description.

Atomic Abstract Query. An atomic abstract query consists of four parts, denoted by {*From, Project, Where, Limit*}. We now describe these components and the way they are computed by function $transTP_m$.

- **From.** The *From* part provides the concrete query that the abstract query relies on. It contains the logical source of an xR2RML mapping, that consists of the `xrr:query` or `rr:tableName` properties, an optional iterator (property `rml:iterator`) and the optional `xrr:uniqueRef` property. With the example of query Q_2 (Listing 1.3), the *From* part for tp_1 simply consists of the logical source of `<#Mbox>: db.people.find({'emails':{$ne: null}})`.
- **Project.** Traditionally, the projection part of a database query restricts the set of attributes that must be returned in the query response. In relational algebra, this is denoted by the projection operator π: $\pi_{a_1,...a_n}(R)$ denotes the tuple obtained when the attributes of tuple R are restricted to the set $\{a_1,...a_n\}$. Similarly, the *Project* part of an atomic abstract query is a set of xR2RML references. For each variable in the triple pattern, the xR2RML references in the term map matched with that variable are projected. In our running example, the subject and object of tp_1 are $?x$ and $?mbox1$. They are matched with the subject and object maps of mapping `<#Mbox>`. Thus, the corresponding xR2RML references within these subject map and object map must be projected. Hence the *Project* part for tp_1: {`$.id AS ?x`, `$.emails.* AS ?mbox1`}. Furthermore, the child and parent joined references of a referencing object map must be projected in order to accommodate databases that do not support joins. In the relational database case, these projections

```
transₘ(gp₂) =
    transTPₘ(tp₁, c₁) INNER JOIN
    transTPₘ(tp₂, true) ON {} INNER JOIN
    transTPₘ(tp₃, c₂) ON {?x,?y}
    FILTER(?x != ?y)

transTPₘ(tp₁, c₁) =
    { From:     {"db.people.find({'emails': {$ne: null}})"},
      Project:  {$.id AS ?x, $.emails.* AS ?mbox1},
      Where:    {isNotNull($.id), isNotNull($.emails.*),
                    sparqlFilter(contains(str(?mbox1),"foo.com"))}}

transTPₘ(tp₂, true) =
    { From:     {"db.people.find({'emails': {$ne: null}})"},
      Project:  {$.id AS ?y},
      Where:    {isNotNull($.id), equals($.emails.*,"john@foo.com")}}

transTPₘ(tp₃, c₂) =
    { From:     {"db.people.find({'contacts':{$size: {$gte:1}}})"},
      Project:  {$.id AS ?x, $.contacts.*},
      Where:    {isNotNull($.id), isNotNull($.contacts.*),
                    sparqlFilter(?x != ?y)}} AS child
    INNER JOIN
    { From:     {"db.people.find({'emails':{$ne: null}})" },
      Project:  {$.emails.*, $.id AS ?y},
      Where:    {isNotNull($.emails.*), isNotNull($.id),
                    sparqlFilter(?x != ?y)}} AS parent
    ON child/$.contacts.* = parent/$.emails.*
```

Listing 1.4. Rewriting of the graph pattern gp_2 of query Q_2 (Listing 1.3) into an abstract query

would be useless since the database can compute the join internally. But the abstract query must accommodate any target database, hence the systematic projection of joined references.

- **Where.** The *Where* part is a set of conditions about xR2RML references. They are produced by matching each term of a triple pattern tp with its corresponding term map in mapping M: the subject of tp is matched with M's subject map, the predicate with M's predicate map and the object with M's object map. Additional conditions are entailed from the SPARQL filter f. In [22], we show that three types of condition may be created:

  (i) a SPARQL variable in the triple pattern is turned into a not-null condition on the xR2RML reference corresponding to that variable in the term map, denoted by *isNotNull(<xR2RML reference>)*;

  (ii) A constant term in the triple pattern (IRI or literal) is turned into an equality condition on the xR2RML reference corresponding to that term in the term map, denoted by *equals(<xR2RML reference>, value)*;

  (iii) A SPARQL filter condition about a SPARQL variable is turned into a filter condition, denoted by *sparqlFilter(<xR2RML reference>, f)*.

Running Example. In the case of query Q_2 (Listing 1.3), triple pattern tp_2 is matched with mapping <#Mbox>. It has the variable ?y in the subject position, which entails an *isNotNull* condition. It also has a constant term in the object

```
transₘ(tp1 AND tp2 AND tp3, c1 ∧ c2) =
  { From:    {"db.people.find({'emails':{$ne:null}})"},
    Project: {$.id AS ?x, $.emails.* AS ?mbox1},
    Where:   {isNotNull($.id), isNotNull($.emails.*),
              sparqlFilter(contains(str(?mbox1),"foo.com"))}}
  INNER JOIN
  { From:    {"db.people.find({'contacts':{$size: {$gte:1}}})"},
    Project: {$.id AS ?x, $.contacts.*},
    Where:   {isNotNull($.id), isNotNull($.contacts.*),
              sparqlFilter(?x != ?y)}} AS child
  INNER JOIN
  { From:    {"db.people.find({'emails':{$ne: null}})"},
    Project: {$.emails.*, $.id AS ?y},
    Where:   {isNotNull($.emails.*), isNotNull($.id),
              equals($.emails.*,"john@foo.com"),
              sparqlFilter(?x != ?y)}} AS parent
  ON child/$.contacts.* = parent/$.emails.* )
  ON {?x,?y}
  FILTER(?x != ?y)
```

Listing 1.5. Optimization of $trans_m(gp_2)$ (Listing 1.4) by self-join elimination

position, which entails an *equals* condition. Finally, the *Where* part for tp_2 contains two conditions: isNotNull($.id) and equals($.emails.*,"john@foo.com"). When we put all the pieces together, we can rewrite the graph pattern gp_2 of SPARQL query Q_2 into the abstract query depicted in Listing 1.4.

4.4   Abstract Query Optimization

At this point, the method we have exposed translates a SPARQL graph pattern into an effective abstract query, *i.e.* that preserves the semantics of the SPARQL query. Yet, shortcomings such as unnecessary complexity or redundancy may lead to the generation of inefficient queries, and consequently yield poor performances. Although we may postpone the query optimization to the translation into a concrete query language, it is beneficial to figure out which optimizations can be done at the abstract query level first, and leave only database-specific optimizations to the subsequent stage.

SPARQL-to-SQL methods proposed various SQL query optimizations such as [14,32,39]. In this section, we review some of these techniques, referring to the terminology defined in [39]. We show how these optimizations can be adapted to fit in the context of our abstract query language. In particular, we show that our translation method implements some of these optimizations by construction. In addition, we propose a new optimization, the *Filter Propagation*, that, to our knowledge, was not proposed in any SPARQL-to-SQL rewriting method.

Filter Optimization. In a naive approach, strings generated by R2RML templates are dealt with using an SQL comparison of the resulting strings rather than the database values used in the template. Typically, when the translation of an R2RML template relies on the SQL string concatenation, a SPARQL query can been rewritten into something like this:

```
SELECT... FROM... WHERE
   ('http://domain/' || TABLE.ID) = 'http://domain/1'
```

Such a query returns the expected results but is likely to perform very poorly: due to the concatenation, the query evaluation engine cannot take advantage of existing database indexes. Conversely, a much more efficient query would be:

```
SELECT ('http://domain/' || TABLE.ID)... FROM...
   WHERE TABLE.ID = 1
```

In our approach, equality conditions apply to xR2RML references rather than on the template-generated values, hence the *Filter Optimization* is enforced by construction.

Filter Pushing. As mentioned earlier, the translation of a SPARQL filter into an encapsulating SELECT WHERE clause lowers the selectivity of inner queries, and the query evaluation process may have to deal with unnecessarily large intermediate results. In our approach, *Filter pushing* is enforced by construction by the *sparqlCond* function: relevant SPARQL conditions are pushed down, as much as possible, in the translation of individual triple patterns.

Self-join Elimination. A self-join may occur when several mappings share the same logical source. This can lead to several triple patterns being translated into atomic abstract queries with the same *From* part. The *Self-join Elimination* consists in merging the criteria of several atomic queries into a single equivalent query. In Listing 1.4, the atomic query in $transTP_m$(tp$_2$, true) and the second atomic query in $transTP_m$(tp$_3$, c$_2$) have the same *From* part and project the same JSONPath expression as variable ?y. Using joins commutativity, those two queries can be merged into a single one depicted in the third atomic abstract query in Listing 1.5[8].

Self-union Elimination. A UNION operator can be created either due to the SPARQL UNION operator or during the translation of a triple pattern to which several mappings are bound (in function $transTP_m$). Analogously to the *Self-join Elimination*, a union of several atomic abstract queries sharing the same logical source can be merged into a single query when they have the same *From* part.

Constant Projection. The *Constant Projection* optimization detects cases where the only projected variables in the SPARQL query are matched with constant values in the bound mappings. In the relational database context, it has been referred to as the *Projection Pushing* optimization [39]. Let us consider the example query below:

```
SELECT DISTINCT ?p WHERE {?s ?p ?o}.
```

In a naive approach, all mappings are bound to the triple pattern ?s ?p ?o. Hence, the resulting abstract query is a union of the atomic queries derived from all the possible mappings. In other words, this query will materialize the whole

---

[8] Note that for a self-join elimination to be safe, additional conditions must be met, that we do not detail here.

```
AND(<exp₁>, <exp₂>, ...)         → $and:[<exp₁>,<exp₂>,...]
OR(<exp₁>, <exp₂>, ...)          → $or:[<exp₁>,<exp₂>,...]
WHERE(<JavaScript exp>)          → $where:'<JavaScript exp>'
ELEMMATCH(<exp₁>,<exp₂>...)      → $elemMatch:{<exp₁>,<exp₂>...}
FIELD(p₁) ... FIELD(pₙ)          → "p₁. ... .pₙ":
SLICE(<exp>, <number>)           → <exp>:{$slice:<number>}
COND(equals(v))                  → $eq:v
COND(isNotNull)                  → $exists:true, $ne:null
EXISTS(<exp>)                    → <exp>:{$exists:true}
NOT_EXISTS(<exp>)                → <exp>:{$exists:false}
COMPARE(<exp>, <op>, <v>)        → <exp>:{<op>:<v>}
NOT_SUPPORTED                    → ∅
UNION(<query1>, <query2>...)     Same semantics as OR, although OR is processed
                                 by the NoSQL engine whereas UNION is processed
                                 by the query processing engine
```

Listing 1.6. Abstract representation of a MongoDB query and translation to a concrete query string. `<op>` stands for one of the MongoDB comparison operators: $eq, $ne, $lt, $lte, $gt, $gte, $size$ and $regex$.

database before it can provide an answer. Very frequently, xR2RML predicate maps are constant-valued: the predicate is not computed from a database value, on the contrary it is defined statically in the mapping. This is typically the case in our running example that has only constant predicate maps (values of property rr:predicate: foaf:knows and foaf:mbox (Listing 1.1). In such cases, given that the SPARQL query retrieves only DISTINCT values of the predicate variable ?p, no query needs to be run against the database at all: it is sufficient to collect the distinct constant values that variable ?p can be matched with. More generally, this optimization checks if the variables projected in the SPARQL query are matched with constant term maps. If this is verified, the SPARQL query is rewritten such that the values of the projected variables be provided as an inline solution sequence using the SPARQL 1.1 VALUES clause. Using the mapping graph of our running example, we would rewrite the query in this way:

```
SELECT DISTINCT ?p WHERE
  { VALUES ?p ( foaf:mbox foaf:knows )}
```

Filter Propagation. We identified another type of optimization that was not implemented in the SPARQL-to-SQL context. This optimization applies to the inner join or left outer join of two atomic queries, and seeks to narrow down one of the joined queries by propagating filter conditions from the other query. In an inner join, if the two queries have shared variables, then *equals* and *isNotNull* conditions of one query on those shared variables can be propagated to the other query. In a left join, propagation can happen only from right to left query since null values must still be allowed in the right query.

5 Application to the MongoDB NoSQL Database

In the previous section, we have exhibited an abstract query model and a method to translate a SPARQL graph pattern into an optimized abstract query, relying on the xR2RML mapping of a target database to RDF. We now want to illustrate the effort it takes to translate from the abstract query language towards a concrete query language with a somewhat different expressiveness.

To this end, we consider the MongoDB NoSQL database. Its JSON-based data model and its query language differ greatly from SQL-based systems for which many rewriting works have been proposed. Hence, we believe that it should provide an interesting illustration of our method. Besides, MongoDB has become a popular NoSQL actor in recent years. It is provided as a service by major cloud service providers and tends to become common within the scientific community, suggesting that it is increasingly adopted as a commonplace database.

In this section, we first glance at the MongoDB query language, and we describe an abstract representation of MongoDB queries (Sect. 5.1). Then, we show that the translation from the abstract query language towards MongoDB is made challenging by the expressiveness discrepancy between the two languages (Sect. 5.2) and we describe a complete method to achieve this. Finally, we summarize the whole SPARQL-to-MongoDB process orchestration, from the SPARQL graph pattern translation until the generation of the RDF triples that match this graph pattern (Sect. 5.3).

5.1   The MongoDB Query Language

MongoDB comes with a rich set of APIs to allow applications to query a database in an imperative way. In addition, the MongoDB interactive interface defines a JSON-based declarative query language consisting of two query methods. The *find* method retrieves documents matching a set of conditions and returns a cursor to the matching documents. Optional modifiers amend the query to impose limits and sort orders. Alternatively, the *aggregate* method allows for the definition of processing pipelines: each document of a collection passes through each stage of a pipeline thereby creating a new collection. This allows for a richer expressiveness but comes with a higher resource consumption that entails less predictable performances. Thus, as a first approach, this work considers the *find* query method, hereafter called the *MongoDB query language*.

The MongoDB *find* query method takes two arguments formatted as JSON documents. The first argument describes conditions about the documents to search for. Query operators are denoted by a heading '$' character. The optional projection argument specifies the fields of the matching documents to return. For instance, the query below matches all documents with a field "emails" and returns only the"id" field of each matching document.

```
db.people.find({"emails": {$exists: true}}, {"id": true})
```

The MongoDB documentation provides a rich description of the *find* query that however lacks precision as to the formal semantics of some operators. Attempts were made to clarify this semantics while underlining some limitations and ambiguities: Botoeva et al. [7] mainly focus on the *aggregate* query and ignore some of the operators we use in our translation, such as *$where*,

$elemMatch$, $regex$ and $size$. On the other hand, Husson [20] describes the *find* query, yet some restrictions on the operator $where$ are not formalized.

Hence, in [22] we specified the grammar of the subset of the query language that we consider. We also defined an abstract representation of MongoDB queries, that allows for handy manipulation during the query construction and optimization phases. Listing 1.6 details the constructs of this representation and their equivalent concrete query string, when relevant. The NOT_SUPPORTED clause helps keep track of any location, within the query, where a condition cannot translate into an equivalent MongoDB query element. It shall be used in the last rewriting and optimization phase.

Let us consider the following abstract representation of a MongoDB query (or "abstract MongoDB query" for short):

```
AND( COMPARE(FIELD(p) FIELD(0), $eq, 10),
     FIELD(q) ELEMMATCH(COND(equals("val")) )
```

It matches all documents where "p" is an array field whose first element (at index 0) is 10, and "q" is an array field in which at least one element has value "val". Its concrete representation is:

```
$and: [ {"p.0": {$eq:10}},
        {"q": {$elemMatch: {$eq:"val"}}} ]
```

5.2 Translation of an Abstract Query into MongoDB Queries

Section 4 elaborated on how a SPARQL graph pattern translates into an abstract query based on xR2RML mappings. Abstract operators INNER JOIN, LEFT OUTER JOIN and UNION relate sub-queries. The lowest level of sub-queries consists of atomic abstract queries of the form {*From, Project, Where, Limit*}, that stem from the translation of individual triple patterns. The *From* part contains the logical source of a mapping bound to the triple pattern to translate. The *Project* part lists the xR2RML data element references that are projected, *i.e.* that are part of the query result. In the context of MongoDB, these xR2RML data element references are JSONPath expressions. The *Where* part is calculated by matching triple pattern terms with relevant xR2RML term maps. This generates conditions on JSONPath expressions (*isNotNull* conditions for SPARQL variables or *equals* conditions for constant triple pattern terms) and *sparqlFilter* conditions that encapsulate SPARQL filters. Finally, the *Limit* part denotes an optional maximum number of results.

Fig. 2. Translation of atomic abstract queries into concrete MongoDB queries

To achieve a translation from the abstract query language towards the MongoDB query language, we must figure out which components of an abstract query have an equivalent MongoDB rewriting, and, conversely, which components shall be computed by the query-processing engine. Below, we analyze the possible situations.

- **Inner and left outer joins.** MongoDB *find* queries do not support joins. Consequently, there does not exist any MongoDB query that would be equivalent to the INNER JOIN and LEFT OUTER JOIN operators. These operators need to be processed by the query-processing engine by joining the RDF triples generated for both sub-queries.
- **UNION.** The rewriting of the UNION operator depends on the graph patterns to which it applies. Let us consider the following SPARQL graph pattern, where tp_n is any triple pattern: {tp₁. tp₂.} UNION {tp₃. tp₄.} Each member of the union translates into an INNER JOIN. Since joins cannot be processed within MongoDB, the outer UNION operator cannot be processed within MongoDB either. The issue occurs likewise as soon as one of the members is either an INNER JOIN or LEFT OUTER JOIN. Under some circumstances, a UNION operator may be translated into the MongoDB *$or* operator. Yet, the MongoDB language definition imposes specific restrictions as to how operators can be nested. Consequently, in a first approach, we always shift the processing of the UNION abstract operator to the query-processing engine. Further works could attempt to characterize more specifically the situations where a UNION can be processed within MongoDB.
- **FILTER and LIMIT.** In Sect. 4, we showed that the FILTER and LIMIT SPARQL solution modifiers are pushed down into relevant atomic abstract queries (as *sparqlFilter* conditions of the *Where* part or as the *Limit* part of an atomic query, respectively). When FILTER and LIMIT SPARQL clauses cannot be pushed down in atomic queries, they end up as abstract operators with the same names, FILTER and LIMIT. The latter apply to abstract sub-queries made of UNION, INNER JOIN and/or LEFT OUTER JOIN operators. Hence, given that UNION and INNER/LEFT OUTER JOIN operators are not processed within MongoDB, the FILTER and LIMIT operators cannot be processed within MongoDB either.

Ultimately, it occurs that only the atomic abstract queries can be processed within MongoDB, while other abstract operators shall be taken care of by the query-processing engine. More generally, the translation from the abstract query language towards MongoDB consists of two steps depicted in Fig. 2. In step 1 (detailed in Sect. 5.2), the translation of each atomic abstract query towards MongoDB amounts to translate projections of JSONPath expressions (*Project* part) into MongoDB projection arguments, and conditions on JSONPath expressions (*Where* part) into equivalent abstract MongoDB queries. Several shortcomings may appear at this stage, such as unnecessary complexity or untranslatable conditions. Thus, in step 2 (detailed in Sect. 5.2) each abstract MongoDB query is optimized and rewritten into valid, concrete MongoDB queries.

In the current status of this work, we do not consider the translation of SPARQL filters (conditions *sparqlFilter*) for the sake of simplicity. SPARQL 1.0

filters come with a broad set of conditional expressions including logical comparisons, literal manipulation expressions (string, numerical, boolean), XPath constructor functions, casting functions for additional data types of the RDF data model, and SPARQL built-in functions (*lang, langmatches, datatype, bound, sameTerm, isIRI, isURI, isBlank, isLiteral, regex*). Handling these expressions within the translation towards MongoDB would yield a significant additional complexity without changing the translation principles though. Yet, an implementation should handle them for the sake of performance and completeness.

Translation of Projections and Conditions. Two functions, named *proj* and *trans*, handle the translation of the *Project* and *Where* parts of an atomic abstract query respectively. Below, we illustrate their principles on an example. The interested reader shall find their formal definition in [22].

In Listing 1.5, the third atomic abstract query is as follows (the *sparqlFilter* condition has been omitted):

```
{From:     {"db.people.find({'emails':{$ne: null}})"},
 Project:  {$.emails.*, $.id AS ?y},
 Where:    {isNotNull($.emails.*), isNotNull($.id),
            equals($.emails.*,"john@foo.com") }}
```

Function *proj* converts the JSONPath expressions of the *From* part into a list of paths to be projected. In the example, expressions `$.emails.*` and `$.id` translate into their MongoDB projection counterparts: `"emails":true` and `"id":true`.

Function *trans* translates a condition of the *Where* part into a MongoDB query element expressed using the abstract representation in Listing 1.6. In the example, condition `isNotNull($.emails.*)` is translated into the following abstract representation:

```
FIELD(emails) ELEMMATCH(COND(isNotNull)).
```

Later on, this abstract representation will be translated into an equivalent concrete query: `"emails": {$elemMatch: {$exists:true, $ne:null}}`. Similarly,   condition   `isNotNull($.id)`   will   be   translated   into: `"id": {$exists:true, $ne:null}`, and condition `equals($.emails.*,"john@foo.com")` will be translated into: `"emails": {$elemMatch: {$eq:'john@foo.com'}}`.

These conditions are used to augment the query of the *From* part, initially provided by the mapping's logical source. When we put all the pieces together, the atomic abstract query is translated into the concrete MongoDB query below, where all conditions are operands of an $and operator:

```
db.people.find(
  # Query argument
  { $and: [
    {"emails": {$ne:null}}, # from the From part
    {"emails": {$elemMatch: {$exists:true,$ne:null}}},
    {"id":    {$exists:true,$ne:null}},
    {"emails": {$elemMatch: {$eq:'john@foo.com'}}} ]
  },
  # Projection argument
  { "emails": true, "id": true }
)
```

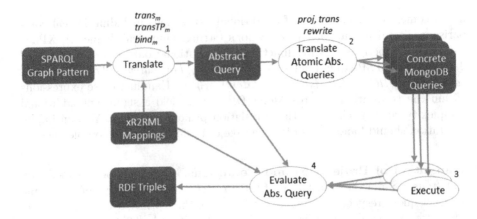

Fig. 3. Complete SPARQL-to-MongoDB query translation and evaluation

Optimization and Rewriting into Concrete MongoDB Queries. In the previous section, function *trans* produces abstract MongoDB queries that can be rewritten into concrete queries straightaway. Yet, this rewriting may be hindered by three potential issues:

(i) During the translation process, nested OR or AND clauses may be produced, as well as sibling WHERE clauses. Such unnecessary complexity may yield an underperforming query.

(ii) It may not be possible to translate some JSONPath expressions into equivalent MongoDB operators. This occurs with specific JSONPath array slice notations, or in JSONPath expressions assuming that the root document is an array field and not a document field (which is forbidden in MongoDB). In such cases, a NOT_SUPPORTED clause tracks the location of this failed translation.

(iii) The MongoDB $where operator passes a JavaScript expression or function to the query system. It provides greater flexibility than other operators, however it is valid only in the top-level query document: it cannot be used inside a nested query such as the $elemMatch operator. During the translation process though, function *trans* may nest a WHERE clause beneath other clauses, yielding an invalid query.

To take care of these issues, in [24] we described a post-translation function *rewrite*, depicted by step 2 in Fig. 2. First, a set of rewriting rules address issue (i) by flattening nested OR, nested AND and nested UNION clauses, and merging sibling WHERE clauses.

To address issue (ii), these rules remove NOT_SUPPORTED clauses while ensuring that the resulting query returns a superset of the valid answers: all the correct answers are returned, along with possibly incorrect answers. In turn, the transformation of this superset into RDF triples shall produce all the triples that match the SPARQL query, in addition to triples that may not match the query.

The latter are ruled out during the query evaluation process by running a late SPARQL query evaluation.

A second set of rewriting rules address issue (iii) by "pulling up" WHERE clauses at the top-level query. This is notably achieved by replacing OR clauses with UNION clauses that have the same semantics but are processed differently. An OR clause represents the $or operator and is processed by MongoDB. Conversely, the UNION clause has no equivalent MongoDB operator: it is processed outside of MongoDB by the query processing engine. As a consequence, an abstract MongoDB query may be rewritten into a union of valid, concrete MongoDB queries.

Finally, Theorem 1 captures two key properties of the rewriting process. It has been proved in [22].

Theorem 1. *Let C be an equality or not-null condition on a JSONPath expression. Let $Q = (Q_1 \ldots Q_n)$ be the abstract MongoDB query produced by trans(C).*
Rewritability: *It is always possible to rewrite Q into a query $Q' =$ UNION$(Q'_1, \ldots, Q'_m)$ such that $\forall i \in [1, m]$ Q'_i is a valid MongoDB query, i.e. Q'_i does not contain any* NOT_SUPPORTED *clause, and a* WHERE *clause only shows at the top-level of Q'_i.*
Completeness: *Executing Q' against the database retrieves all the documents matching condition C. If Q contains at least one* NOT_SUPPORTED *clause, then Q' may retrieve additional documents that do not match condition C.*

A corollary of Theorem 1 is that, using the xR2RML mapping of a MongoDB database to RDF, we can rewrite any SPARQL 1.0 graph pattern into an abstract query whose atomic abstract queries are valid MongoDB queries or unions of valid MongoDB queries.

5.3 Complete SPARQL-to-MongoDB Query Translation and Evaluation

Figure 3 summarizes the whole SPARQL-to-MongoDB process orchestration, from the graph pattern translation to the subsequent MongoDB queries evaluation and the production of RDF triples.

In step 1, function $trans_m$ (Sect. 4.1) translates a SPARQL graph pattern into an abstract query under a set of xR2RML mappings denoted by m. It leverages function $transTP_m$ (Sect. 4.3) to translate a triple pattern tp into an abstract query under the set of mappings bound to tp by function $bind_m$ (Sect. 4.2). The resulting abstract query contains atomic abstract queries of the form {*From, Project, Where, Limit*}, combined with abstract operators INNER JOIN, LEFT OUTER JOIN, UNION, FILTER, LIMIT. The *Project* part of an atomic abstract query is a set of xR2RML references (*i.e.* JSONPath expressions for MongoDB) that must be projected. The *Where* part consists of *isNotNull*, *equals* and *sparqlFilter* conditions on JSONPath expressions. In step 2, function *proj* translates each projected JSONPath expression into a MongoDB projection argument, function *trans* translates each *isNotNull* and *equals* condition into an abstract representation of a MongoDB query (Sect. 5.2), and function *rewrite* (Sect. 5.2) optimizes and rewrites this abstract representation into a concrete MongoDB query or a union of concrete MongoDB queries.

Two steps remain, that we have not described yet. In step 3, the concrete queries are executed against the database. In step 4, the result JSON documents are translated into RDF triples according to the xR2RML mappings, then the query processing engine evaluates the abstract query by computing the INNER/LEFT OUTER JOIN, UNION, FILTER and LIMIT operators. Finally, in case one atomic abstract query contained a NOT_SUPPORTED clause, a late SPARQL evaluation is performed to rule out the RDF triples that do not match the query (as explained in Sect. 5.2).

6   Experimentation and Evaluation

To date, to our knowledge, the method proposed in this paper and the MongoDB-enabled *ontop* software [8] are the only approaches meant to query arbitrary MongoDB documents with SPARQL. So far though, this *ontop* version is not available for test, which hinders possible performance comparison. Additionally, no benchmark similar to the Berlin SPARQL Benchmark for relational databases [6] exists so far for querying NoSQL databases with SPARQL.

Therefore, in this section, we describe a real-world use case that we used to build a test database, and we report experimental results with respect to the effectiveness and performance of our approach.

6.1   Prototype Implementation

Morph-xR2RML is the prototype implementation we developed to evaluate the effectiveness of the xR2RML mapping language and the SPARQL-to-MongoDB method proposed in this paper. It comes with connectors for the MySQL and Postgres relational databases, and for the MongoDB document store. It can process an xR2RML mapping graph in either the data materialization or the query rewriting modes.

Morph-xR2RML is available on GitHub[9] under the Apache 2.0 license, it is written in the Scala programming language. It is based on and extends the Morph-RDB [29] R2RML implementation. We performed a substantial code refactoring in order to isolate any RDB-related code into a dedicated software module. As a result, our prototype is extensible by design: supporting a new type of database amounts to create a new software module that implements a given set of interfaces, thereby encapsulating and isolating any database-specific concerns from the rest of the project code. Following this approach, we developed a connector for the MongoDB document store, to translate MongoDB JSON documents into RDF and rewrite SPARQL queries into MongoDB queries.

---

[9] https://github.com/frmichel/morph-xr2rml/.

Morph-xR2RML relies on several open-source Java APIs, the most salient ones are listed here. Jena[10] is a well known Java framework consisting of several APIs meant to build Semantic Web Data applications. We use the Jena RDF API that helps handle RDF triples and graphs. MongoDB comes with a native Java API[11] that allows for imperative style querying only. The Jongo API[12] builds on top of it to translate a declarative MongoDB query (a *find* query in our case) into imperative code. Lastly, Jayway JsonPath[13] is a Java implementation of the JSONPath language.

The query rewriting experiment we report in this section was conducted on a server equipped with a 3.0 GHz CPU with two physical cores, and 8 GB RAM. The MongoDB engine and the Morph-xR2RML Java virtual machine alike were running on the same server. The Java virtual machine was allowed a maximum of 4 GB memory.

6.2   Experimentation Database

TAXREF [15] is the French national taxonomic register for fauna, flora and fungus, maintained and distributed by the French National Museum of Natural History (MNHN). It is a manually curated register of all the species inventoried in metropolitan France and overseas territories, organized as a hierarchy of over 485.000 scientific names (in version 9) that mark a national and international consensus. As an example, the listing below shows a JSON excerpt from TAXREF's Web service[14], describing the common dolphin species (*Delphinus delphis*). Annotation `"habitat":1` states that it lives in a marine habitat, annotation `"rang":"ES"` states that the taxon belongs to the "species" taxonomical rank. Annotation `"fr":"P"` characterizes one of its biogeographical statuses: it states that *Delphinus delphis* is present in mainland France.

```
{
  "codeTaxon":"60878",
  "codeReference":"60878", "codeParent":"191591",
  "rang":"ES",
  "libelleNom":"Delphinus delphis",
  "libelleAuteur":"Linnaeus, 1758",
  "nomVernaculaire":"Dauphin commun",
  "nomVernaculaireAnglais":"Common Dolphin",
  "url":"http://inpn.mnhn.fr/espece/cd_nom/60878",
  "habitat":"1",
  "fr":"P",
  (...)
}
```

We are involved in an on-going collaboration with TAXREF experts from MNHN, aimed to publish TAXREF on the Web of Data as a SKOS thesaurus [9]. In this context, we imported into a MongoDB database the JSON representation of TAXREF v9.0, wherein each of the 485.189 JSON documents accounts for

---

[10] http://jena.apache.org/.

[11] https://mongodb.github.io/mongo-java-driver/.

[12] http://jongo.org/.

[13] https://github.com/json-path/JsonPath.

[14] https://taxref.mnhn.fr/taxref-web/api/doc.

one scientific name, may it be a taxon reference of synonymous name. Listing 1.7 exemplifies the SKOS modeling with taxon *Delphinus delphis* and its synonym *Delphinus vulgaris*. The taxon is represented as a SKOS concept (line 10). The `skos:broader` property models the relationships towards the parent taxon in the classification (line 13), *i.e.* genus *Delphinus* in this example. The taxon reference and synonymous names are represented as SKOS-XL labels (lines 23–33), referred to with properties `skosxl:prefLabel` and `skosxl:altLabel` respectively (lines 14–15). The taxonomical rank, habitat and bio-geographical status are properties of the SKOS concept (lines 16–21), while the authorities and vernacular names are properties of SKOS labels (lines 25–27 and 31–33).

Leveraging this existing database, we set up an experimentation of the SPARQL-to-MongoDB query rewriting. In the next section, we shortly describe the xR2RML mappings designed for the experimentation.

```
1   @prefix txrp:     <http://inpn.mnhn.fr/taxref/properties/> .
2   @prefix txrbgs:   <http://inpn.mnhn.fr/taxref/bioGeoStatus#>.
3   @prefix nt:       <http://purl.obolibrary.org/obo/ncbitaxon#>.
4   @prefix dwc:      <http://rs.tdwg.org/dwc/terms/>.
5   @prefix txn:      <http://lod.taxonconcept.org/ontology/txn.owl#>.
6   @prefix dct:      <http://purl.org/dc/elements/1.1/>.
7   @prefix skos:     <http://www.w3.org/2004/02/skos/core#>.
8   @prefix skosxl:   <http://www.w3.org/2008/05/skos-xl#>.
9
10  <http://inpn.mnhn.fr/taxref/9.0/taxon/60878> a skos:Concept;
11    skos:inScheme        <http://inpn.mnhn.fr/taxref/9.0/Taxref>;
12    skos:note            "Delphinus delphis";
13    skos:broader         <http://inpn.mnhn.fr/taxref/9.0/taxon/191591>;
14    skosxl:prefLabel     <http://inpn.mnhn.fr/taxref/label/60878>;
15    skosxl:altLabel      <http://inpn.mnhn.fr/taxref/label/577834>;
16    txrp:habitat         <http://inpn.mnhn.fr/taxref/habitat#Marine>;
17    nt:has_rank          <http://inpn.mnhn.fr/taxref/taxrank#Species>;
18    txrp:bioGeoStatusIn  [
19      rdfs:label            "Metropolitan France";
20      dct:spatial           <http://sws.geonames.org/3017382/>;
21      dwc:locationId        "TDWG:FRA; WOEID:23424819";
22      dwc:occurrenceStatus  txrbgs:P ].
23
24  <http://inpn.mnhn.fr/taxref/label/60878> a skosxl:Label;
25    txrp:isPrefLabelOf   <http://inpn.mnhn.fr/taxref/9.0/taxon/60878>;
26    txn:authority        "Linnaeus, 1758";
27    txrp:vernacularName  "Common Dolphin"@en, "Dauphin commun"@fr;
28    skosxl:literalForm   "Delphinus delphis".
29
30  <http://inpn.mnhn.fr/taxref/label/577834> a skosxl:Label;
31    txrp:isAltLabelOf    <http://inpn.mnhn.fr/taxref/9.0/taxon/60878>;
32    txn:authority        "Lacepede, 1804";
33    txrp:vernacularName  "Common Dolphin"@en, "Dauphin commun"@fr;
34    skosxl:literalForm   "Delphinus vulgaris".
```

Listing 1.7. SKOS representation of the *Delpinus delphis* taxon

6.3 Experimentation xR2RML Mapping Graph

The xR2RML mapping graph designed to generate the TAXREF-based SKOS thesaurus is provided in the xR2RML GitHub repository[15]. It consists of 90

---

[15] xR2RML mapping graph for TAXREF v9: https://github.com/frmichel/morph-xr2rml/blob/master/morph-xr2rml-dist/example_taxref/xr2rml_taxref_v9.ttl.

mappings, a somewhat high number that spawns from the distance between the internal structure of TAXREF JSON documents and the targeted SKOS modeling. We illustrate this distance with an example.

Habitats are coded in TAXREF with integer values, *e.g.* value '1' represents the marine habitat, '2' represents fresh water, etc. Translating the marine habitat into URI http://inpn.mnhn.fr/taxref/habitat#1 would be straightforward using a template that would append the value read from the database to http://inpn.mnhn.fr/taxref/habitat#. A single mapping would be sufficient to generate the triples related to all types of habitat. However, our modeling targets the generation of more meaningful URIs that cannot be generated by a template, e.g. http://inpn.mnhn.fr/taxref/habitat#Marine; instead, we must write a mapping whose query filters only taxa with habitat '1':

```
<#TM_Habitat_Marine>
 xrr:logicalSource [ xrr:query """db.taxrefv9.find(
    {$where: 'this.codeTaxon==this.codeReference',
     'habitat':'1'} )""" ];
 rr:subjectMap <#SM_Taxon>;
 rr:predicateObjectMap [
  rr:predicate txrfp:habitat;
  rr:objectMap [
   rr:constant
    <http://inpn.mnhn.fr/taxref/habitat#Marine>;
   rr:termType rr:IRI ]].
```

Such a mapping must be written for each of the 8 habitat values. A similar situation is observed for the 48 taxonomical ranks and 30 bio-geographical statuses, that all comme with dedicated mappings.

6.4   Experimentation Results

In Sect. 5, we showed that atomic abstract queries can be translated into equivalent MongoDB queries, but other operators of the abstract query language (INNER JOIN, LEFT OUTER JOIN, UNION) must be computed by the query-processing engine, *i.e.* Morph-xR2RML. Therefore, a first series of tests aimed to assess the performance of Morph-xR2RML with a SPARQL query consisting of a single triple pattern, bound to exactly one mapping and producing a single MongoDB query (Sect. 6.4). In a second series of tests, we measured the completion time of SPARQL queries involving joins and/or unions, and we compared them to the time needed for a single triple pattern. Furthermore, we measured the gain obtained by performing optimizations at the level of the abstract query (Sect. 6.4).

Processing a Single Triple Pattern. To measure the performance of Morph-xR2RML in the case of a single triple pattern translated into a single MongoDB query, we selected seven SPARQL SELECT queries (Q0 to Q6) tailored to produce an increasing number of results: from 1 result in Q0 to 227,224 results in Q6. In each case, one JSON document yields one RDF triple. Table 1 lists each query along with the corresponding triple pattern and semantics, the number of results it retrieves from the database, and the average time it took to process

Table 1. Execution time of SPARQL queries with one triple pattern

Q. Id	Query semantics and SPARQL triple pattern	No. results	Exec. time ± std dev. (ms)	Exec. time per result (ms)
Q0	*Find the reference name for taxon 60587* ?t skosxl:prefLabel <http://inpn.mnhn.fr/taxref/label/60587>	1	451 ± 36	451.00
Q1	*Get synonyms of taxon 95372* <http://inpn.mnhn.fr/taxref/9.0/taxon/95372> skosxl:altLabel ?a	164	522 ± 14	3.18
Q2	*Get all bio-geographical statuses in* *St Pierre et Miquelon* ?bgs dct:spatial <http://sws.geonames.org/3424932/>	4835	4.056 ± 65	0.84
Q3	*Get all bio-geographical statuses in Guadeloupe* ?bgs dct:spatial <http://sws.geonames.org/3579143/>	17956	9665 ± 45	0.54
Q4	*Get all bio-geographical statuses in* *New Caledonia* ?bgs dct:spatial <http://sws.geonames.org/2139685/>	35703	17289 ± 78	0.48
Q5	*Get bio-geographical statuses in mainland France* ?bgs dct:spatial <http://sws.geonames.org/3017382/>	128018	61645 ± 671	0.48
Q6	*Get all taxa (that are SKOS concepts)* ?c a skos:Concept	227224	108508 ± 459	0.48

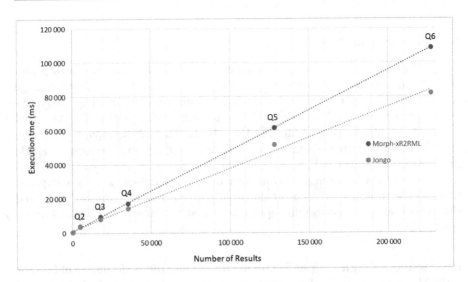

Fig. 4. Average query processing time as a function of the number of results. Dotted lines represent the linear regression lines of both series. (Color figure online)

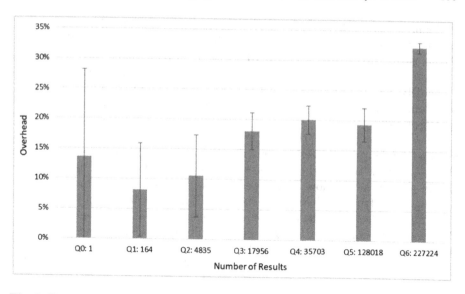

Fig. 5. Processing time overhead imposed by Morph-xR2RML, compared to a direct database query. The overhead comprises rewriting the SPARQL query and translating the MongoDB results into RDF triples

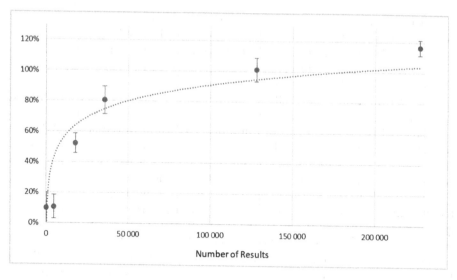

Fig. 6. Overhead of querying MongoDB through the Jongo API compared to a direct query through MongoDB's Java API

the query (the query processing spans the SPARQL query rewriting, the query evaluation against MongoDB and the RDF triples generation). For each query, 10 measures were performed: we report the average value and standard deviation. The last column gives the average processing time per query result, that converges towards 0.48 ms.

Figure 4 depicts the average query processing time (fourth column of Table 1) as a function of the number of results (blue line). Since Morph-xR2RML relies on the Jongo API to process a MongoDB query, we also measured the time needed by Jongo to parse the query, pass it on to MongoDB and retrieve the results from MongoDB. Red dots represent the measures when simply querying MongoDB with Jongo, while blue dots represent the measures of the whole process executed by Morph-xR2RML.

The distance between the two lines gives an estimation of the overhead imposed by Morph-xR2RML to rewrite the query and generate the triples. Figure 5 depicts this overhead. The confidence for Q0 and Q1, and to some extend for Q2, is very low as attested by the large error bars. Indeed, materializing a few triples is barely measurable (<1 ms for Q0, and in the order of 30 ms for Q1), such that the measure is very sensitive to environment variations. Conversely, the confidence for Q3 to Q6 is quite high. Q3, Q4 and Q5 show a similar overhead of approximately 19%. Although we could expect the overhead percentage to be constant with higher numbers of results, it reaches 32% for Q6. A detailed analysis shows that the difference lies in the time needed to generate the RDF triples. Compared to Q5, the number of results in Q6 increases by 77% while the materialization time increases by 120%. The variable term in Q3, Q4 and Q5 is a blank node whereas it is a URI in Q6. A tentative explanation is that Morph-xR2RML may be faster when producing blank nodes than when producing URIs, unless this difference lies in the Jena API on which Morph-xR2RML relies to handle RDF triples. Further works should consider using more substantial databases to assess this difference with more precision. In any case, the processing performed by Morph-xR2RML adds no more than a 30% overhead to the time needed to query the database and retrieve the results.

Yet, waiting 10 s to get 18000 results (query Q3) can be considered surprisingly long compared to native RDF triple stores. To investigate this question, we compared the time it takes to run a query (i) through the Jongo API (the case of Morph-xR2RML) and (ii) directly through MongoDB's own Java API. The results are presented in Fig. 6. Surprisingly, they attest that, while Jongo is efficient for few results (in the order of 100), it entails a significant overhead for larger results: 116% overhead for query Q6 (*i.e.* using Jongo more than doubles the query time). Jongo's authors argue that the library is almost as fast as querying MongoDB directly, under the assumption that the marshalling/unmarshalling of JSON documents is left to Jongo. Morph-xR2RML retrieves JSON documents from Jongo as Java strings in order to evaluate them with JSON-Path expressions. It is likely that converting documents to strings and evaluating them with a third-party JSONPath library significantly impairs performances. Further investigation should be conducted to figure this out more precisely, keeping in mind that solving this issue could approximately save a factor 2 during the processing of large result sets.

Impact of Query Optimizations. In this section, we measure the completion time of two example SPARQL queries involving joins. Notably, we measure the gain obtained by performing optimizations at the level of the abstract query, namely the self-join elimination and the filter propagation. Additional example queries are reported in [22] along with measures of the impact of the self-union elimination and the constant projection optimizations.

Join Query, Self-join Elimination. SPARQL query Q_7, depicted below, looks for taxa (variable *?t*) that are present in the overseas collectivity of Saint-Pierre-et-Miquelon (http://sws.geonames.org/3424932/). The graph pattern matches 12,708 triples that yield a SPARQL result set of 4,236 solutions.

```
SELECT * WHERE {
    ?t taxrefprop:bioGeoStatusIn ?bgs .          # tp1
    ?bgs dct:spatial
        <http://sws.geonames.org/3424932/> .      # tp2
    ?bgs dwc:occurrenceStatus taxrefbgs:P .       # tp3
}
```

Executed separately, the first triple pattern would be bound to 15 mappings (one for each geographical location) and would yield 311,489 RDF triples; the second one would be bound to one mapping and would yield 4,835 triples, and the third one would be bound to 15 mapping and would yield 260,631 documents. Executed as such, query Q_7 completes in almost 10 min (600 s).

```
[ { Binding(tp1: ?t taxrefprop:bioGeoStatusIn ?bgs -> TM_SBG_SPM)
    From    : db.taxrefv9.find({$where:'this.codeTaxon==this.codeReference',
                                'spm':{$ne:''},'spm':{$ne:null}})
    Project: $.codeTaxon AS ?t, $.codeTaxon AS ?bgs
    Where  : isNotNull($.codeTaxon) }
] INNER JOIN [
    [ { Binding(tp2: ?bgs dct:spatial http://sws.geonames.org/3424932/
                      -> TM_SBG_SPM_BN2)
        From
    : db.taxrefv9.find({$where:'this.codeTaxon==this.codeReference',
                        'spm':{$ne:''}, 'spm':{$ne:null}})
        Project: $.codeTaxon AS ?bgs
        Where  : isNotNull($.codeTaxon) }
    ] INNER JOIN [
        { Binding(tp3: ?bgs dwc:occurrenceStatus taxrefbgs:P ->TM_SBG_SPM_BN1)
          From
    : db.taxrefv9.find({$where:'this.codeTaxon==this.codeReference',
                        'spm':{$ne:''}, 'spm':{$ne:null}})
        Project: $.codeTaxon AS ?bgs)
        Where  : isNotNull($.codeTaxon), equals($.spm, P) }
    ] ON ?bgs
] ON ?bgs


{ Binding(tp1: ?t    taxrefprop:bioGeoStatusIn ?bgs -> TM_SBG_SPM),
  Binding(tp2: ?bgs dct:spatial http://sws.geonames.org/3424932/
                    -> TM_SBG_SPM_BN2),
  Binding(tp3: ?bgs dwc:occurrenceStatus taxrefbgs:P -> TM_SBG_SPM_BN1)
  From   : db.taxrefv9.find({$where:'this.codeTaxon == this.codeReference',
                            'spm':{$ne:''}, 'spm':{$ne:null}})
  Project: $.codeTaxon AS ?t, $.codeTaxon AS ?bgs
  Where  : isNotNull($.codeTaxon), equals($.spm, P)
}
```

Listing 1.8. Top: rewriting of the graph pattern of query Q_7 after bindings reduction. Bottom: the same query after self-join elimination.

Compared to the notation used in previous sections, each atomic abstract query contains heading lines providing the binding(s) of the triple pattern(s) that this atomic query accounts for, denoted by `Binding(triple pattern -> mapping name)`.

```
[{ Binding(?t skosxl:prefLabel http://inpn.mnhn.fr/taxref/label/60585
                 -> TM_Taxon_PrefLabel)
    From    : db.taxrefv9.find( {
                  $where: 'this.codeTaxon==this.codeReference' } )
    Project: $.codeTaxon AS ?t
    Where   : isNotNull($.codeTaxon), equals($.codeTaxon, 60585) }
] INNER JOIN [
    [{ Binding(?t skosxl:altLabel ?a -> TM_Taxon_AltLabel)
        From    : db.taxrefv9.find( {
                      $where: 'this.codeTaxon!=this.codeReference' } )
        Project: $.codeReference AS ?t, $.codeTaxon AS ?a
        Where   : isNotNull($.codeReference), isNotNull($.codeTaxon) }
    ] INNER JOIN [
      { Binding(?t skosxl:altLabel ?b -> TM_Taxon_AltLabel)
        From    : db.taxrefv9.find( {
                      $where: 'this.codeTaxon!=this.codeReference' } )
        Project: $.codeReference AS ?t, $.codeTaxon AS ?b
        Where   : isNotNull($.codeReference), isNotNull($.codeTaxon) }
    ] ON ?t
] ON ?t
```

Listing 1.9. Rewriting of the graph pattern of query Q_8.

The binding reduction step (Sect. 4.2) removes all but one mapping bound to the first and third triple patterns. The query now amounts to the join of three atomic abstract queries depicted in Listing 1.8 (top). The first and second atomic queries yield 4,835 RDF triples while the third query yields 4,236 triples. Under such reduced bindings, query Q_7 completes in 8.53 s in average, the querying to MongoDB accounts for 47% of this total time, the generation of the RDF triples accounts for 11% and the processing of joins for 39%.

A closer look to the abstract query shows that it contains two self-joins that can be eliminated for the following reasons: (i) all three queries share the same *From* part (the logical source), (ii) they are joined on the *?bgs* variable that is always projected from the same reference $.codeTaxon, and (iii) $.codeTaxon is declared as a unique identifier in at least one mapping bound to the three triple patterns (with property xrr:uniqueRef). This self-join elimination yields an optimized query that now consists of a single atomic query depicted in Listing 1.9 (bottom). Note that the *Project* and *Where* parts have been merged, and the three bindings now apply to this atomic query: the same MongoDB query is used to generate RDF triples matching the three triple patterns. This optimized query completes in 2,966 ms in average, *i.e.* a 65% gain compared to the query with reduced bindings.

Filter Propagation. SPARQL query Q_8, pictured herebelow, retrieves the taxon (variable *?t*) whose preferred label has a certain URI, alongside two of its alternate labels (variables *?a* and *?b*).

```
SELECT * WHERE {
  ?t skosxl:prefLabel
     <http://inpn.mnhn.fr/taxref/label/60585> .
  ?t skosxl:altLabel ?a .
  ?t skosxl:altLabel ?b .
  FILTER (?a != ?b)
}
```

In a first step, Q_8 translates into the inner join of three atomic abstract queries, portrayed in Listing 1.9. The first atomic query retrieves 1 document from the database, while the second and third queries retrieve 257,965 documents each. Executed naively, the inner-most join computes the join of 257,965 triples with another 257,965 triples generated from the same database documents. With a smarter join ordering, the triple produced by the first atomic query is joined with the 257,965 triples of the second one to produce two triples (taxon 60585 has two synonyms), that, in turn, are joined with the 257,965 triples of the third query. Yet, two joins of 257,965 triples with one then two triples have to be performed. Some tests show that the time needed to complete this query is in the order of 4 min.

```
[{ Binding(?t skosxl:prefLabel http://inpn.mnhn.fr/taxref/label/60585
                -> TM_Taxon_PrefLabel)
    From    : db.taxrefv9.find( {
                $where: 'this.codeTaxon==this.codeReference' } )
    Project: $.codeTaxon AS ?t
    Where   : isNotNull($.codeTaxon), equals($.codeTaxon, 60585) }
] INNER JOIN [
    [{ Binding(?t skosxl:altLabel ?a -> TM_Taxon_AltLabel)
        From    : db.taxrefv9.find( {
                    $where: 'this.codeTaxon!=this.codeReference' } )
        Project: $.codeReference AS ?t, $.codeTaxon AS ?a
        Where   : isNotNull($.codeTaxon), equals($.codeReference, 60585) }
    ] INNER JOIN [
      { Binding(?t skosxl:altLabel ?b -> TM_Taxon_AltLabel)
        From    : db.taxrefv9.find( {
                    $where: 'this.codeTaxon!=this.codeReference' } )
        Project: $.codeReference AS ?t, $.codeTaxon AS ?b
        Where   : isNotNull($.codeTaxon), equals($.codeReference, 60585) }
    ] ON ?t
] ON ?t
```

Listing 1.10. Rewriting of the graph pattern of query Q_8 after enforcing the filter propagation optimization.

The Filter Propagation optimization leverages some situations where, within the join of two sub-queries, a condition on a variable shared by both sub-queries can be propagated from one sub-query to the other. In the example, the two joins are performed on variable ?t. The first atomic query projects ?t as expression $.codeTaxon and has condition equals($.codeTaxon, 60585). In the second and third queries, variable ?t is projected as $.codeReference. Therefore, the join condition can only be satisfied if expression $.codeReference returns the value 60585. In other words, we can propagate the condition on $.codeTaxon, equals($.codeTaxon, 60585) to the second and third queries as a condition on $.codeReference: equals($.codeReference, 60585). The optimized abstract query is pictured in Listing 1.10. The second and third queries now only yield two RDF triples. Finally, the execution of this query lasts 565 ms in average, that is a gain factor in the order of 400.

7 Discussion and Perspectives

In the case of MongoDB, the processing of joins is shifted to the query processing engine, and can ensue poor performances when joined sub-queries are not selective enough. Furthermore, real-world SPARQL queries often contain substantial graph patterns with multiple joined triple patterns. It is therefore critical to be able to process joins efficiently. Thus, beyond the optimizations that we implemented at the abstract query level, query-plan optimization techniques shall be investigated to help answer the following questions:

- Can we rewrite a SPARQL graph pattern in a way that facilitates the production of an efficient abstract query?
- How to inject intermediate results into a subsequent query, as performed in the bind join optimization [17]?
- How to reorder joins considering the number of results of sub-queries, in a way similar to methods proposed by distributed query engines? [16,21,33]
- Can we perform lazy evaluation of joins by progressively materializing triples on each side of the join until the expected number of results is reached? This would typically resemble the method employed in the non-blocking evaluation of queries in the context of Triple Pattern Fragments [40].

Additionally, several leads could be investigated to overcome the limitations of the translation from the abstract query language to MongoDB.

- Our method generates the RDF triples resulting from each atomic queries and subsequently performs joins (INNER JOIN, LEFT OUTER JOIN). In some cases though, joins may rule out many of the triples that were just materialized. Hence, it should be studied when joins can be evaluated on the database documents. This would typically rule out unnecessary documents earlier in the process, thus saving the useless generation of RDF triples.
- Our implementation of xR2RML for MongoDB relies on JSONPath to extract data elements from MongoDB results. In turn, the SPARQL rewriting process must handle conditions on JSONPath expressions. Consequently, we have to cope with the expressiveness discrepancy between SPARQL and MongoDB, and between JSONPath and MongoDB alike. While we must cope with the earlier (our goal is specifically to access heterogeneous databases with SPARQL), the latter is somewhat more an implementation choice. Hence, an investigation should figure out whether considering a restricted subset of JSONPath may produce a simpler solution while still enabling to address most mapping situations.
- Beyond this, another promising lead is to determine what type of MongoDB query should be used preferably: *find* or *aggregate* queries. We address this question in Sect. 7, as part of a broader discussion about the similarities and discrepancies between our approach and that of *ontop*'s authors.

Comparison with the MongoDB-Enabled *Ontop*. To the best of our knowledge, the only other approach meant to access arbitrary MongoDB documents with SPARQL has been proposed by the authors of *ontop*, Botoeva et al. [8]. This approach starts with deriving a set of type constraints (literal, object, array) from the mapping assertions, called the MongoDB database schema. Then, a relational view over the database is defined with respect to that schema, notably by flattening array fields. A SPARQL query is rewritten into relational algebra (RA) query, and RA expressions over the relational view are translated into MongoDB *aggregate* queries. Similarly, we translate a SPARQL query into an abstract representation (that is not relational algebra) under xR2RML mappings. To deal with the tree structure of JSON documents we use JSONPath expressions. On the one hand, this avoids the definition of a relational view over the database, but this comes with additional complexity in the translation process, as translating conditions on JSONPath expressions is not straightforward. On the other hand, the advantage of our method is that the query evaluation relies on existing database indexes, whereas in the case of Botoeva et al., the flattening step prevents from exploiting these indexes.

The mappings are quite similar in both approaches although xR2RML is more flexible: (i) class names (in triples `?x rdf:type A`) and predicates can be built from database values whereas they are constant in the approach proposed by Botoeva et al., and (ii) xR2RML allows to turn an array field into an RDF collection or container, while the latter approach only supports the multiple-triples strategy.

Finally, the main differences pertain to the type of target query. Botoeva et al. produce MongoDB *aggregate* queries, with the major advantage of ensuring a semantics-preserving SPARQL-to-MongoDB query translation, thus delegating the whole processing to MongoDB and making the query translation simpler. In practice however, *aggregate* pipelines may perform poorly. To optimize them, an option suggested by the authors is to decompose the pipeline into smaller queries and have the query-processing engine perform the remaining steps. Our approach works the other way around: it produces less-expressive MongoDB *find* queries, leaving much more work to the query-processing engine. Nevertheless, having the job done outside of the database engine allows to leverage extensive works about smart query optimizations [16,17,21,33], whereas this is not possible when the database performs an *aggregate* query in a black-box manner.

Typically though, in situations involving large joins, *aggregate* queries perform faster than *find* queries as they can leverage database indexes. In the future, it would be interesting to assess whether we could characterize mappings with respect to the type of query that shall perform best: single vs. multiple separate queries, *find* vs. *aggregate*, and figure out a balance between the two approaches.

Furthermore, unlike *ontop*, xR2RML allows for rich JSONPath expressions to evaluate a JSON document and generate RDF terms. In this matter, further studies should figure out how to translate such expressions into *aggregate* queries.

8 Conclusion

The method proposed in this paper aims at fostering the development of
SPARQL interfaces to heterogeneous databases, as we believe this is a key to
push the Web of Data forward. In particular, we think that this should help to
bridge the gap between the Semantic Web and the NoSQL family of databases.

To achieve this goal without defining yet another SPARQL translation
method for each and every database, we proposed a two-phase approach. First,
we defined an abstract query language deriving from the syntax and semantics
of SPARQL. Utilizing the xR2RML mapping language and leveraging R2RML-
based SPARQL-to-SQL works, we introduced a generic method to translate a
SPARQL 1.0 graph pattern into an abstract query. We showed how optimizations can be beneficially enforced at this abstract level, saving subsequent work
at the level of a target database language. In a second phase, the abstract query
is translated into the query language of a target database. To demonstrate the
effectiveness of our approach, we applied it to the MongoDB NoSQL document
store. We devised a method to translate an abstract query into MongoDB *find*
queries, and we showed that this translation is challenged by the expressiveness
discrepancy between SPARQL and the MongoDB query language.

Finally, we conducted an experimentation based on the real-world use case of
a taxonomical reference stored in a MongoDB database. Utilizing a mapping of
this database to a SKOS thesaurus, we first measured performances in the case
of single SPARQL triple patterns that translate into single MongoDB queries.
Then, we measured the performances of richer SPARQL queries and we demonstrated the effectiveness of some of the optimizations performed at the level of
the abstract query language. We underlined some limitations of the translation
from the abstract query language to MongoDB, that can impair performances.
In Sect. 7 we discuss several improvement leads that could be investigated.

From a broader perspective, we have shown that translating a SPARQL query
into efficient concrete queries can be challenging when it comes to address data
sources such as NoSQL databases. These systems are generally optimized for fast
storage and retrieval of vast collections of documents. They favor scalability, high
throughput and availability over consistency and query language expressiveness.
As a consequence, they often come with denormalized data models where redundancy is common, and barely support joins. This is the case of other document
stores such as CouchDB that are designed in a way very similar to MongoDB.
Column family stores usually allow for a richer data model and provide a more
expressive query language. But although their columnar data model makes them
easily compared with relational systems, they often suffer the same limitations
as document stores with respect to the limited support of joins. Key-value stores
are designed for fast retrieval of data *e.g.* accessed by key. They are typically used
to implement cache systems, for which a very simple query language (consisting
essentially of *put* and *retrieve* by key operations) covers most use cases.

Consequently, it is likely that the hurdles we encountered with MongoDB will
be encountered with other NoSQL databases alike. The situation may not be so
much different for the last category of NoSQL databases, namely graph stores.

By nature, their data models are closer to RDF. Still, whereas RDF predicates can be used with literal values as well as resources, graph databases such as Neo4J[16] manage literals (called node attributes) and other graph nodes in a very different way. As a result, querying a graph database with SPARQL may be more challenging that it seems, and we believe that our two-phase approach may be relevant in this context too.

References

1. Arenas, M., Bertails, A., Prud'hommeaux, E., Sequeda, J.: A Direct Mapping of Relational Data to RDF (2012)
2. Berners-Lee, T.: Linked Data, in Design Issues of the WWW (2006). http://www.w3.org/DesignIssues/LinkedData.html
3. Bikakis, N., Tsinaraki, C., Gioldasis, N., Stavrakantonakis, I., Christodoulakis, S.: The XML and Semantic Web Worlds: Technologies, Interoperability and Integration: a Survey of the State of the Art. In: Anagnostopoulos, I., Bieliková, M., Mylonas, P., Tsapatsoulis, N. (eds.) Semantic Hyper/Multimedia Adaptation. SCI, pp. 319–360. Springer, Heidelberg (2013). https://doi.org/10.1007/978-3-642-28977-4_12
4. Bikakis, N., Tsinaraki, C., Stavrakantonakis, I., Gioldasis, N., Christodoulakis, S.: The SPARQL2XQuery interoperability framework. World Wide Web 18(2), 403–490 (2015)
5. Bizer, C., Cyganiak, R.: D2R server - publishing relational databases on the semantic web. In: Proceeding of the 5th International Semantic Web Conference (ISWC) (2006)
6. Bizer, C., Schultz, A.: The Berlin SPARQL benchmark. Int. J. Semant. Web Inf. Syst. 5(2), 1–24 (2009)
7. Botoeva, E., Calvanese, D., Cogrel, B., Rezk, M., Xiao, G.: A formal presentation of MongoDB (extended version) (2016). https://arxiv.org/abs/1603.09291v1
8. Botoeva, E., Calvanese, D., Cogrel, B., Rezk, M., Xiao, G.: OBDA beyond relational DBs: a study for MongoDB. In: Proceedings of the 29th International Workshop on Description Logics (2016)
9. Callou, C., Michel, F., Faron-Zucker, C., Martin, C., Montagnat, J.: Towards a shared reference thesaurus for studies on history of zoology, archaeozoology and conservation biology. In: Semantic Web For Scientific Heritage (SW4SH), ESWC Workshops (2015)
10. Chebotko, A., Lu, S., Fotouhi, F.: Semantics preserving SPARQL-to-SQL translation. Data Knowl. Eng. 68(10), 973–1000 (2009)
11. Cyganiak, R., Wood, D., Lanthaler, M.: RDF 1.1 Concepts and Abstract Syntax. W3C Recommendation (2014)
12. Das, S., Sundara, S., Cyganiak, R.: R2RML: RDB to RDF mapping language. W3C Recommendation (2012)
13. Dimou, A., Vander Sande, M., Colpaert, P., Verborgh, R., Mannens, E., Van de Walle, R.: RML: a generic language for integrated RDF mappings of heterogeneous data. In: Proceedings of the 7th Workshop on Linked Data on the Web (2014)
14. Elliott, B., Cheng, E., Thomas-Ogbuji, C., Ozsoyoglu, Z.M.: A complete translation from SPARQL into efficient SQL. In: Proceedings of the International Database Engineering and Applications Symposium, pp. 31–42. ACM (2009)

---

[16] Neo4J: https://neo4j.com/.

15. Gargominy, P., et al.: TAXREF v9. 0, référentiel taxonomique pour la France: Méthodologie, mise en oeuvre et diffusion
16. Görlitz, O., Staab, S.: SPLENDID: SPARQL endpoint federation exploiting VOID descriptions. In: International Workshop on COLD (2011)
17. Haas, L., Kossmann, D., Wimmers, E., Yang, J.: Optimizing queries across diverse data sources. In: Proceedings of the 23rd International Conference on Very Large Data Bases (VLDB 1997), pp. 276–285 (1997)
18. Harris, S., Seaborne, A.: SPARQL 1.1 Query Language. W3C Recommendation (2013)
19. Heath, T., Bizer, C.: Linked Data: Evolving the Web into a Global Data Space, 1st edn. Morgan & Claypool, San Rafael (2011)
20. Husson, A.: Une sémantique statique pour MongoDB. In: 25th Journées Francophones des Langages Applicatifs, pp. 77–92 (2014)
21. Macina, A., Montagnat, J., Corby, O.: Optimising SPARQL query processing in distributed knowledge graphs. In: Actes de la Conférence Gestion de Données - Principes, Technologies et Applications (BDA). Poitiers, France (2016)
22. Michel, F.: Integrating Heterogeneous Data Sources in the Web of Data. Ph.d. thesis, Université Côte d'Azur, March 2017
23. Michel, F., Faron-Zucker, C., Montagnat, J.: A generic mapping-based query translation from SPARQL to various target database query languages. In: Proceeding of the 12th International Conference on Web Information Systems and Technologies (WebIST), vol. 2, pp. 147–158 (2016)
24. Michel, F., Faron-Zucker, C., Montagnat, J.: A mapping-based method to query MongoDB documents with SPARQL. In: Hartmann, S., Ma, H. (eds.) DEXA 2016. LNCS, vol. 9828, pp. 52–67. Springer, Cham (2016). https://doi.org/10.1007/978-3-319-44406-2_6
25. Michel, F., Djimenou, L., Faron-Zucker, C., Montagnat, J.: Translation of heterogeneous databases into RDF, and application to the construction of a SKOS taxonomical reference. In: Monfort, V., Krempels, K.-H., Majchrzak, T.A., Turk, Ž. (eds.) WEBIST 2015. LNBIP, vol. 246, pp. 275–296. Springer, Cham (2016). https://doi.org/10.1007/978-3-319-30996-5_14
26. Mugnier, M.L., Rousset, M.C., Ulliana, F.: Ontology-mediated queries for NOSQL databases. In: Proceedings of the 30th Conference on Artificial Intelligence. Phoenix, Arizona (2016)
27. Pérez, J., Arenas, M., Gutierrez, C.: Semantics and complexity of SPARQL. ACM Trans. Database Syst. **34**(3), 1–45 (2009)
28. Pollock, R., Tennison, J., Kellogg, G., Herman, I.: Metadata Vocabulary for Tabular Data. W3C Recommendation (2015)
29. Priyatna, F., Corcho, O., Sequeda, J.: Formalisation and experiences of R2RML-based SPARQL to SQL query translation using Morph. In: Proceeding of the World Wide Web Conference (WWW) (2014)
30. Rodríguez-Muro, M., Calvanese, D.: High performance query answering over DL-Lite ontologies. In: Proceedings of the 13th International Conference on Principles of Knowledge Representation and Reasoning (KR 2012) (2012)
31. Rodríguez-Muro, M., Kontchakov, R., Zakharyaschev, M.: Ontology-based data access: *Ontop* of databases. In: Alani, H., et al. (eds.) ISWC 2013. LNCS, vol. 8218, pp. 558–573. Springer, Heidelberg (2013). https://doi.org/10.1007/978-3-642-41335-3_35
32. Rodríguez-Muro, M., Rezk, M.: Efficient SPARQL-to-SQL with R2RML mappings. Web Semant. **33**, 141–169 (2015)

33. Schwarte, A., Haase, P., Hose, K., Schenkel, R., Schmidt, M.: FedX: optimization techniques for federated query processing on linked data. In: Aroyo, L., et al. (eds.) ISWC 2011. LNCS, vol. 7031, pp. 601–616. Springer, Heidelberg (2011). https://doi.org/10.1007/978-3-642-25073-6_38

34. Sequeda, J., Tirmizi, S.H., Corcho, O., Miranker, D.P.: Survey of directly mapping SQL databases to the semantic web. Knowl. Eng. Rev. **26**(4), 445–486 (2011)

35. Sequeda, J.F., Miranker, D.P.: Ultrawrap: SPARQL execution on relational data. Web Semant. **22**, 19–39 (2013)

36. Spanos, D.E., Stavrou, P., Mitrou, N.: Bringing relational databases into the semantic web: a survey. Semant. Web J. **3**(2), 169–209 (2012)

37. Tomaszuk, D.: Document-oriented triplestore based on RDF/JSON. In: Logic, Philosophy and Computer Science, pp. 125–140. University of Bialystok (2010)

38. Unbehauen, J., Stadler, C., Auer, S.: Accessing relational data on the web with SparqlMap. In: Takeda, H., Qu, Y., Mizoguchi, R., Kitamura, Y. (eds.) JIST 2012. LNCS, vol. 7774, pp. 65–80. Springer, Heidelberg (2013). https://doi.org/10.1007/978-3-642-37996-3_5

39. Unbehauen, J., Stadler, C., Auer, S.: Optimizing SPARQL-to-SQL rewriting. In: Proceedings of Information Integration and Web-based Applications & Services (iiWAS 2013), p. 324. ACM (2013)

40. Verborgh, R., et al.: Triple pattern fragments: a low-cost knowledge graph interface for the web. Web Semant. **37–38**, 184–206 (2016)

Author Index

Printed in the United States
By Bookmasters